Environmental Movements of India

Environmental Movements of India

Chipko, Narmada Bachao Andolan, Navdanya

Krishna Mallick

Routledge
Taylor & Francis Group
LONDON AND NEW YORK

First published in 2021 by Amsterdam University Press Ltd.

Published 2025 by Routledge
4 Park Square, Milton Park, Abingdon, Oxon OX14 4RN
605 Third Avenue, New York, NY 10158

Routledge is an imprint of the Taylor & Francis Group, an informa business

ISBN: 9789462984431 (hbk)
ISBN: 9781041178828 (pbk)
ISBN: 9781003694861 (ebk)
NUR 740

Cover photo: Fran Smith

Cover design: Coördesign, Leiden

DOI 10.5117/9789462984431

For Product Safety Concerns and Information please contact our EU representative:
GPSR@taylorandfrancis.com
Taylor & Francis Verlag GmbH, Kaufingerstraße 24, 80331 München, Germany

Dedicated to my granddaughter, Meera,
who is growing up in an environmentally challenging world

Table of Contents

Acknowledgments 11

Foreword 13
Frances Moore Lappé

Glossary 15

Introduction 17
Three Grassroots Movements That Made a Global Impact
Principles of Environmental Philosophy 20
(1) Environmental Justice 20
(2) Intergenerational Equality 20
(3) Respect for Nature 21

1 Historical and Cultural Contexts in India 23
1.1 'Legal' Destruction of India's Forests 24
1.1.1 Acts Spark Peasant Protests 25
1.2 A Cultural Leader Emerges 27
1.2.1 Nonviolence and Gandhi's Truth 28
1.2.2 How Chipko Followed Hinduism through Gandhi 29
1.2.3 How NBA Followed Hinduism through Gandhi 30
1.2.4 How Navdanya Followed Hinduism through Gandhi 31
1.2.5 Ecology and Social Justice 31
1.3 Conclusion 33

2 Chipko (Hug the Trees) Movement 37
2.1 A Physical Act of Survival 38
2.1.1 Sacred Texts and Social Justice 39
2.2 Preconditions and Formation of the Chipko Movement 40
2.2.1 Workers Organize for Nonviolent Action 42
2.2.2 Suffering by Means of Fasting and Foot March 43
2.3 Laudable Leaders 44
2.3.1 Women's Role in the Chipko Movement 46
2.4 Critical Reception of the Chipko Movement 48
2.4.1 Questions about Chipko's Popularity and Success 52
2.5 Conclusion 55

3 **Narmada Bachao Andolan (NBA): Save the Narmada** 59
3.1 The Common Good in a Cost-Benefit Analysis 61
3.1.1 A Recursive History of Dam-Building 63
3.2 Regional Tensions from the Start 64
3.2.1 Amid Unrest, NGOs Align to Form NBA 65
3.2.2 National and Global Ramifications 67
3.3 Gender and the Narmada Case 68
3.3.1 Roles for Displaced Women 71
3.4 Gendered Dimensions of Neoliberal Capitalist Development 73
3.5 Reasons for the Success of the NBA 75
3.6 Conclusion 77

4 **Navdanya (Nine Seeds) Movement** 83
4.1 The Terrible Human Toll of GM Crops 84
4.1.1 Shifting Economics 86
4.2 Emergence of Anti-GM Movements 87
4.2.1 KRRS: Fiery Fields of Protest 87
4.2.2 Gene Campaign: Secure Food and Climate 88
4.2.3 Navdanya's Holistic Approach 89
4.3 Food Sovereignty 90
4.3.1 Biodiverse Organic Farming 92
4.4 Biodiversity and Climate Change 93
4.4.1 Entrepreneurial Renewal 95
4.5 Navdanya and Social Justice 96
4.5.1 Civil Disobedience 97
4.5.2 Human Right to Food 98
4.5.3 Protecting the Global South 101
4.6 Shiva's View of Earth Democracy 103
4.7 Genetically Modified Crops and the Future 106
4.8 Conclusion 109

5 **Moral Implications of Environmental Movements** 115
5.1 The Mesmerizing Power of Nonviolence 115
5.1.1 An Ecological Warrior 117
5.1.2 Truth at All Costs 119
5.2 Defining Views of Globalism 119
5.2.1 Technological Prowess 120
5.3 Core Values of Development Ethics 121
5.3.1 Environmental Justice for All 130

5.4 Ecofeminism: Ethics of Mutual Care and Connection 131
5.4.1 Ecofeminist Roots in the Chipko Movement 133
5.4.2 Southern Materialist Ecofeminism 135
5.5 Conclusion 136

6 Hindu Ethics and Ecology 143
6.1 Historical Background of Hinduism 145
6.2 Comparison of Hindu *Dharma* and Ethics in the West 147
6.2.1 The Gita and Dharma 150
6.2.2 The Ramayana and Dharma 152
6.2.3 The Yogasutra and Dharma 153
6.3 Hindu *Dharma,* Ecology, and Sustainability 153
6.3.1 Hindu *Dharma* and Applications in Ecologically Sustainable Development 159
6.4 Ways Hindus Connect to Nature 161
6.5 Influence of Symbolic Traditions on Some Environmental Cases 163
6.6 Is Hinduism Eco-Friendly? 166
6.7 Influence of Hinduism and Other Literature on Gandhi 168
6.8 Conclusion 170

Conclusion 175
The Symbiosis of Natural Resources and Local Needs
Theoretical Views of the Global South 177
Global Environmental Theory 177
Feminist Care Ethics 178
The Capabilities Approach 179
Ethics of Nonviolence 179

Index 183

Acknowledgments

The idea of writing a book about the grassroots environmental movements of India had occurred to me for a long time, since I have been teaching the course *Environmental Ethics* for many years. In teaching this course, I have always devoted the last two or three classes to some of the grassroots environmental movements of India. I searched many times for a supplementary textbook but couldn't find one that covered some of these movements. It is this desire to have a book that led me to write the book proposal.

I am grateful to my colleague, Dr. Severin Kitanov in the Philosophy Department at Salem State University, who helped me to bring this book to fruition. Dr. Kitanov read my entire manuscript and gave me extensive comments to improve the quality of the content. My heartfelt thanks to him for the many hours he spent reading and commenting at every phase of the book.

I want to thank my students, from whom I have learned so much over the course of many years. I also came to know about their interests in and curiosity about environmental issues in India, as they don't hear or read about them in the mainstream media.

During the initial period of writing this book, Prof. Patricia Ould, a colleague in the Sociology Department at my institution, made comments that were vital in clarifying many ideas of the book. I must also thank Dr. Monica V. Ogra, Associate Professor at Gettysburg College, who took many hours from her busy life to read chapters and offer valuable comments. I thank Laura Smith for her tireless and thorough work of editing and polishing the final manuscript.

I thank my reviewer, Dr. Somnath Bandyopadhyay of Nalanda University, whose suggestions for the benefit of environmental studies students in India have been invaluable. I want to thank my second reviewer for her helpful comments.

Thanks to several mini-grants from Salem State University that helped me with this project.

Above all, I am grateful to my husband, Subir, for his encouragement and patience.

Foreword

Frances Moore Lappé

I promise that you will feel better after reading this book. Bringing to life underappreciated social movements that have made a huge, positive difference in our world, *Environmental Movements of India* is a gift. It is needed now more than ever. Krishna Mallick's profound passion for global environmental justice shines throughout as it brings to life key environmental movements in India.

Dr. Mallick's expertise in both Eastern and Western philosophy along with her background in environmental studies, women's studies, and peace studies bring a unique perspective to our understanding of the sources, range, and impact of environmental activism.

The three environmental movements of India that Dr. Mallick explores have developed over the last 50 years and reveal significant, fascinating differences between Eastern environmentalism and its Western counterpart. In an era obsessed with economic growth and dangerously blind to environmental sustainability, this book shows how the voice of the disadvantaged is key in reducing social inequities that are worsening worldwide.

Environmental Movements of India gives voice to those who are not in a position of authority and power but who have still made an impact not only at the national level in India but also on an international and global scale.

Dr. Mallick's chosen case studies – the Chipko movement, the NBA, and Navdanya – demonstrate the profound inspiration of strong, Indian environmental ethics that have sustained environmental movements over many decades. At the same time, they showcase the potential of such ethics to provide principles for an alternative framework for global struggles in environmental conservation. A critical contribution of the author is bringing to light their roots in a philosophical framework that ranges from the heritage of Mahatma Gandhi to various environmental movements manifesting enduring environmental ethics. Furthermore, her analysis reflects a deep familiarity with the relevant literature on the debate about environmental vs. economic development.

The history of the Chipko, NBA, and Navdanya movements matters not only because of what these movements have accomplished in India. The leaders and followers of the movements have also embodied the ideals of the Gandhian method of *satyagraha*. When peoples' very survival and livelihood were threatened by destructive economic forces, Gandhian *satyagraha* offered a powerful model of resistance against injustice and oppression. The Indian government was eventually forced to come to terms with the demands of the movements, which, in turn, made a global impact.

The Western world can thus learn valuable lessons from the experience of these Indian environmental movements. Most importantly, Dr. Mallick has clearly demonstrated that ethics and morality cannot be ignored simply because no law has been violated. The protection of human beings and nature must always be the centerpiece of making economic policy.

The stories of the leaders and followers of the three movements call out to readers to do much more to ensure that all stakeholders – in India and across all nations – are present at the table and to acknowledge the primacy of environmental sustainability and social justice as well as the necessity of treating those who suffer with the dignity that we all deserve.

Glossary

Note the different translations of Sanskrit terms.

achit	non-sentient matter
adivasi	indigenous people or tribal people or scheduled tribe
ahimsa	non-violence, non-injury
anna swaraj	food sovereignty
aparigraha	non-possessiveness
artha	prosperity, accumulation of wealth (one of the aims of life)
asteya	non-stealing
atman	higher self
Beej Bachao Andolan	Seed Freedom Movement
Bhagavad Gita or *Gita*	Song of God
bhakti	devotion
bhakti yoga	way of devotion
bij swaraj	seed sovereignty
brahmacharya	chastity
Brahman	the Absolute
Brahmin	ascetic or priest caste
chipko	cling on to or hold on to (hug the trees)
chit	sentient matter
darshan	to see or vision, Sanskrit word for philosophy
dharma	righteousness or duty or moral code or social morality; literal meaning 'to uphold' or 'sustain' (one of the aims of life)
Dharmashastras	treatises on dharma
gram swaraj	village self-reliance
gunas	attributes of nature or matter
Ishwarapranidhana	Meditation of *Ishwara* or God
jal swaraj	water sovereignty
jnana yoga	way of knowledge
kama	fulfillment of desire, sensual desire (one of the aims of life)
karma yoga	way of action
kshatriya	warrior caste
khadi	homespun cotton

kuladharma	duty of the family
moksha	inner freedom or liberation or salvation (one of the aims of life)
navdanya	nine crops or seeds; the word 'nav' means new and the word 'daan' means gift or offering well-being of all creatures
nishkama karma	doing one's action without thinking about the consequences or doing one's action with detachment or non-attachment
niyama	discipline
padayatra	foot march
prakriti	nature or matter
purusarthas	four aims of life
purusha	the soul or self
rajas	energy or activity or being passionate (this is a *guna*)
rakhees	wrist bands used during festivals
santosha	contentment
sarvodaya	welfare for all, universal uplift
sattva	purity or wisdom or lightness (this is a *guna*)
satya	truthfulness
satyagraha	truth-force or soul-force
saucha	cleanliness
swadeshi	honouring the local, homegrown
swadharma or samanyadharma	common duty, duties that everyone is required to follow
swadhyaya	habit of study or reflection
swaraj	self-rule, self-discipline
tamas	heaviness or darkness or inert or resistant (this is a *guna*)
tapas	stoic endurance of discomfort without complaint
tapasya	austerity, self-discipline
van panchayats	community forestry management
varnadharma	duty of the caste
yama	restraint

Introduction

Three Grassroots Movements That Made a Global Impact

Abstract
Three Indian environmental movements are analyzed – the Chipko movement of the 1970s against deforestation, Narmada Bachao Andolan in the 1980s against dam-building, and Navdanya, the contemporary movement against genetically modified organisms (GMOs) – from three dimensions: nonviolence, feminism, and environmentalism. Each of these movements has accomplished its goals, with Chipko achieving a 15-year ban on tree-cutting, NBA succeeding in slowing down the building of dams on the Narmada river through litigation with the help of the World Bank and still fighting for the resettlement and rehabilitation of the displaced people, and Navdanya promoting local organic seeds for small farmers in India whose rights have been violated by multinational corporations monopolizing GM crops. The three movements followed three principles: environmental justice, intergenerational equality, and respect for nature.

Keywords: environmental justice, intergenerational equality, respect for nature, globalization from below, globalization from above

Many people believe, and rightly so, that the main source of environmental destruction in the world is the demand for natural resources caused by consumption by the rich, whether nations, individuals, or groups. The positive message of this book is that poor communities have the power to accomplish economically and environmentally sound goals. In various parts of India over the last five decades, several social justice and environmental movements have been formed and continue to emerge. Environmental sustainability has been the primary goal of many of these movements, and my aim in this book is to provide detailed accounts of three such movements that had both a local and a global impact.

Mallick, Krishna, *Environmental Movements of India: Chipko, Narmada Bachao Andolan, Navdanya*. Taylor & Francis Group, 2021.
DOI: 10.5117/9789462984431_INTRO

The Chipko movement, with its focus on preventing deforestation; Narmada Bachao Andolan (NBA), which fought for the rights of villagers displaced by a dam; and the Navdanya movement, a seed repository of native varieties, together constitute a special triad in terms of the magnitude of their impact, their membership constituencies, and their lack of political affiliation. This book honours their contributions toward an ethics of 'globalization from below' as opposed to 'globalization from above'. In simple terms, globalization from below is the struggle against social exclusion that involves coming up with new initiatives that have both local and global impacts. By contrast, globalization from above – neoliberalism – helps to maximize capitalist efficiency and economic growth through free markets and policies that promote multinational companies' (deleterious) involvement in the resources of other countries.

The Chipko, NBA, and Navdanya movements significantly changed the development discourse by including not only people in power but also the disempowered and disadvantaged villagers whose lives were directly impacted by decisions made without their input. Each movement was led by one or two organizers and was followed by scores of people who protested the actions of the central and state Indian governments as well as the multinational and transnational companies that continue to coordinate with private industries in India with the help of the government. Generally speaking, these three grassroots movements have exercised significant pressure on the Indian government to modify or change its policies related to deforestation, dam-building, and food sovereignty.

The guiding purpose of my analysis is to present and make understandable for a variety of readers the voice of a developing country such as India, a voice that is very different from that of developed nations in general. More precisely, my aim is to help readers appreciate the nonviolent, feminist, and environmental dimensions of these three environmental movements. In each, the Gandhian principle of *satyagraha* – soul-force or truth-force – is employed in various ways. *Satyagraha* originates from the belief that while violence to persons and property diverts the minds of the parties from the real issues at hand, nonviolent action invites the parties to a dialogue about the issues themselves. Nonviolence (*ahimsa*) is not just the absence of violence (*himsa*). At the core of *satyagraha* is a commitment to the truth of a situation, surrounded by the suffering needed to induce a change of heart in the oppressor. A practitioner of nonviolence must be disciplined to face the inevitable suffering and possibility of death and must not use any form of violence in thought, speech, or action.

The preconditions for India's environmental movements are made clear in Chapter 1 with an overview of the numerous Forest Acts enforced by the British Raj before and briefly after India's independence. These Acts led to peasant protests and the emergence of a cultural leader in Mohandas K. Gandhi, who had returned to India after practicing law for two decades in South Africa, where he had been honoured with the name Mahatma. In the 1970s, Chipko activists were concerned with issues of social justice arising from the disproportionate use of forest resources by people in positions of authority and power. As described in Chapter 2, the movement became iconic worldwide for the nonviolent action taken by local women of hugging trees that had been marked for felling. Directly following independence, Prime Minister Jawaharlal Nehru launched the Sardar Sarovar dam project as part of a vision of mass industrialization. Chapter 3 tells the story of Narmada Bachao Andolan (NBA), an anti-dam group that emerged in the 1980s. A pivotal moment in its history came when the NBA convinced the World Bank to withdraw its funding for the dam. Yet due to the powerlessness of displaced people, the harm done has been significant and complex.

Amid the debate over genetically modified organisms (GMOs), Chapter 4 begins with a look back at the Green Revolution of the 1960s and 1970s. The resulting monoculture crops have eliminated the choices available to local small farmers. Following an estimated 300,000 farmer suicides, movements arose to challenge development models based on efficiency alone. Chapter 4 highlights the work of Navdanya, a seed repository that works directly with local farmers to promote biodiverse organic crops and the concept of food sovereignty. Chapter 5 presents the philosophical and moral underpinnings of environmental movements. I delve further into Gandhi's philosophy of nonviolence, consider the forces of globalism, and examine the six core values of development ethics. As each of the movements covered in this volume demonstrates, ecofeminist principles offer a path away from neoliberalism and toward meeting the basic needs of local people by providing them with access to their natural resources. The final chapter makes the case for Hindu ethics of *dharma* being the foundation of Hinduism and its relation to the sacredness of nature as depicted in many Hindu scriptures. An integration of the eco-friendly nature of Hinduism with environmental awareness is necessary in this Anthropocene era. Concluding that the symbiosis of natural resources and local needs is the ethical goal, I re-examine the three movements through the theoretical lenses of global environment ethics, feminist care ethics, and the capabilities approach.

Five decades later, all three movements remain relevant. The spirit of Chipko is alive in a global movement to ensure women's sovereign rights to natural forest resources. In August 2019, NBA founder Medha Patkar and six supporters participated in a hunger strike to compel authorities to open the sluice gates of the Sardar Sarovar Dam in Gujarat. The Navdanya seed repository, founded by eco-activist Vandana Shiva, is thriving in India and has been adopted as a model by developing nations worldwide.

Principles of Environmental Philosophy

As the discipline of environmental studies is not limited to any specific academic field, this work can be characterized as interdisciplinary, at the intersection of women's studies, environmental studies, peace studies, philosophy, and the emerging field of ecological economics. Specialists in any of these fields as well as general readers will find valuable lessons and insights here, as each of the environmental movements is brought into sharp relief by means of an extensive review of the critical literature. Because people's voices are value-oriented, it is also instructive to identify and explicate the normative principles underpinning these movements, as Tongjin Yang argues in his 2006 article 'Towards an Egalitarian Global Environmental Ethics'.

(1) Environmental Justice

Different studies have shown that the cost of environmental degradation is largely borne by the economically disadvantaged people both within and across countries. The principle of environmental justice – and its related outcome, social justice – demands that the powerless and disadvantaged in society change the injustices caused by the people in power in a nonviolent way, compelling them to move in a positive direction.

(2) Intergenerational Equality

Every future generation deserves the basic right to a healthy and livable environment. This means that every present generation has the responsibility or duty to leave the earth in a state that is politically and economically just as well as sustainable for future generations to be able to live with basic rights.

(3) Respect for Nature

The integrity and biodiversity of nature must be preserved at any cost. As human beings are part of nature, it is essential for them to follow the laws of nature. Respect for nature means that humans should not take away society's resources to give themselves an economic advantage. This requires a sea change from a discourse on economic development to one on sustainability, as the Earth is in crisis now. We must, therefore, carry out our duty to care for it.

1 Historical and Cultural Contexts in India

Abstract
This chapter explores the historical background of British rule in India (1757-1947) and the different Forest Acts that were imposed by the colonial authorities that infringed the rights of the people. This eventually led to protests against the forest management, most of which were suppressed by the British government. Community forest management councils (*van panchayat*) were formed to protect the inhabitants but then the British classified most of the forest as reserve forests, centralizing their control over the forest and displacing the people. Gandhi's method of nonviolence, *satyagraha*, is analyzed by showing how Gandhi was influenced by the *Gita*, Tolstoy, Ruskin, and Thoreau and how Gandhian views have been followed in the Chipko, NBA, and Navdanya movements.

Keywords: *satyagraha*, Indian Forest Acts, *van panchayat*, Gandhi, Hinduism

Over the long period of British rule in India (1757-1947), economic and political power became concentrated in the East India Company, whose shares were owned by wealthy merchants and aristocrats. The East India Company had operated in India from the beginning of British colonialism, but a watershed moment in its history occurred in 1772 when, in order to avoid bankruptcy, the company requested loans from the Bank of England. British rule under the Crown began in 1858, replacing the East India Company's rule. Later on, the East India Company was dissolved and the British Raj was established, which dominated the Indian subcontinent for almost another century.

'British colonialism and other forms of Western imperialism in general have profoundly altered the world's ecology, contributing to far-reaching environmental degradation', Nayak (2015, p. 252) writes. As Nayak points out,

Mallick, Krishna, *Environmental Movements of India: Chipko, Narmada Bachao Andolan, Navdanya*. Taylor & Francis Group, 2021.
DOI: 10.5117/9789462984431_CH01

the ideology of economic development that took root during the colonial period imposed a superstructure of ideas, institutions, and a legal rationale for acquiring control over the entire forested realm of Indian territory. By 1860, Nayak says, 'Britain had emerged as the world leader in deforestation, devastating its own woodlands as well as the forests of Ireland, South Africa, and the northeastern United States' (p. 253). Timber was used for shipbuilding, iron ore for melting, and large tracts of land for cash-crop farming. With oak forests vanishing in England, the Royal Navy had to look elsewhere for a permanent supply of durable timber, for the safety of the empire depended on its wooden ships (Nayak, 2015).

1.1 'Legal' Destruction of India's Forests

The British Parliament passed numerous legislative acts to lay claim to the natural resources of India. The main objectives were to acquire land for railways; to expand trade routes and convert forests and pastureland into plantations of tea, coffee, rubber, and indigo (a tropical tree in the pea family that is used for blue dye); to establish army barracks and ordinance factories; and to promote the construction of dams, canals, etc. Those who were negatively affected in the process of securing the aforementioned objectives were mostly the forest-dwelling communities, tribal communities, and farmers (Fernandes & Paranjpya, 1997; Nayak, 2015). For example, many women's livelihood depended on collecting firewood from the forest to use for cooking for the family. The Indian Forest Act of 1865 made it easier for the wooded areas to be acquired, thereby taking away the existing rights and legitimate interests of people in India's forests and completely restricting the local communities' access to forest resources, which essentially deprived them of their principal source of livelihood (Saxena, 2008; Nayak, 2015). The Forest Act was replaced 13 years later by a far more comprehensive law known as the Indian Forest Act of 1878, which differentiated forests into three classes: reserved forests, protected forests, and village forests. In effect, it 'provided the underpinnings for the "scientific" management of India's forests, enabling the working of compact blocks of forest for commercial timber production' (Nayak, 2015, p. 254).

The imperial government would continue to use India's forests even during the First World War. The British government further declared an act known as the Land Acquisition Act of 1894. Under this act, the colonial government was given the power and authority to acquire private land for public purposes (Nayak, 2015). The Indian Forest Act of 1865 and the Land

Acquisition Act of 1894 established absolute state control over the forests and promoted the dispossession and conversion of land into state property in the name of securing the public good. Consequently, the communities that had depended on forests for centuries lost their customary rights and source of livelihood (Nayak, 2015).

Even though the Forest Law was implemented during colonial times, it is still enforced in present-day India. As Nayak (2015) explains, 'Colonial policies have continued in post-independence India. State developmental projects such as large-scale dam projects, mining, and industries have created the main conditions that have shaped the environmental movements of India' (p. 250). By imposing control over woodlands traditionally controlled by local communities and by using forests for commercial timber production, the colonial state interfered in the day-to-day life of Indian villages. More precisely, by promoting the planting of commercial species (teak, pine, and deodar) in different ecological zones, the state altered a stable forest ecology that had endured for centuries. The uprooted and replaced species (for example, oak and Terminalia) had always been used for fuel, fodder, leaf manure, and small timber. The newly planted species were invariably of little use to the rural population.

Furthermore, the colonial state radically redefined property rights, imposing a system of forest management at odds with the previously existing democratic structure of local use and control. Thus, colonial forestry and land acquisition caused the rise and intensification of conflict over the use of and access to forestland. As more and more people began to protest the policies of the colonial government, various movements emerged in different parts of the country (Gadgil & Guha, 2003; Guha & Gadgil, 1989; Nayak, 2015).

1.1.1 Acts Spark Peasant Protests

The Indian Forest Act of 1878, in particular, can be seen as the major factor behind the rise of countryside protests and movements. A variety of organizations raised their voices against this act. The Poona Sarvajanik Sabha, a nationalist organization in western India, vehemently opposed the Forest Act in 1878. Although this organization primarily represented the Indian middle class, it consistently fought for the rights of the rural cultivators. Triggered by the callous disregard of village needs, demands, and legitimate interests, major peasant protests in response to the Forest Act occurred in 1904, 1906, 1916, 1921, 1930, and 1942. However, with the exception of the movement in the Kumaon and Garhwal regions in 1921, all these movements were suppressed by the British colonial regime (Guha, 1999, 2001).

The 1921 movement in Kumaon and Garhwal was actually the most significant and influential forest protest movement during the colonial regime. The movement was occasioned by the export of timber (deodar, chir, fir, and spruce) as a strategically valuable resource in Britain's military campaign in World War I. The export of timber continued to benefit Britain during World War II as well (Guha, 2001; Nayak, 2015). After the popular uprising in 1921, the government of the United Provinces set up the Kumaon Forest Grievances Committee. The committee conducted a survey of the area and concluded that any attempt to enforce the forest rules would lead to riots and bloodshed. The analysis of the committee prompted the British to divide the existing reserved forests into two main categories: Class I and Class II forests. Class I forests were to be managed by the local civil administration, which was more sympathetic to rural needs, while Class II forests, which included all commercially valuable wooded areas, were to be managed directly by the Forest Department (Nayak, 2015).

A policy passed in 1930 allowed for the formation of *van panchayats,* or community forestry management, in Kumaon and Garhwal. These *van panchayats* covered an area of only less than half a million hectares (or 1235526.91 acres). The implementation of the *van panchayat* in Kumaon and Garhwal constituted the sole network of village forest areas mandated by law in India, but this management structure worked remarkably well with respect to both forest protection and cultivation (Nayak, 2015). Although forests were initially placed on the State List of the Constitution of India, Nayak notes that the 42nd Amendment Act of the Constitution transferred the country's forest resources from the State List to the so-called Concurrent List, which gave both the central and state governments authority to make laws relating to India's forests (Nayak, 2015). The Forest Act of 1927 had been enacted for the purpose of protecting forest habitat, but what it really did was establish government control and create a monopoly over the use of forests with the concept of Reserved Forests. The ongoing process of administrative centralization by consecutive governments further increased the number of people displaced and contributed to the people's alienation from their lands and traditional sources of livelihood.

Sadly, the inherited colonial paradigm of forest appropriation and associated tribal exploitation has been terminologically repackaged in post-independence India as a modern 'development' and has been ideologically projected as a national aspiration. This paradigm has been extended and implemented through a host of different projects, schemes, and programmes, but the ideological underpinnings have remained the same. Forest Acts implemented since the time of India's independence have not changed much in the status quo (Nayak, 2015).

1.2 A Cultural Leader Emerges

This was the tense political landscape to which Mahatma Gandhi returned in 1914, after having worked for two decades as a lawyer and activist in South Africa. He immediately began to acquaint himself with the economic and social conditions of the villages to which he travelled. In 1917 and 1919, Gandhi organized peasant protests in the districts of Champaran and Kheda. His deeper understanding of the roots of colonialism helped him realize that it would be impossible for India to emulate Western patterns of industrial development.

While Gandhi was not an environmentalist, Eliza Kent says that he believed in the unity of all beings, as demonstrated by his vegetarianism and his work fighting for disempowered people in society for most of his life. A Vaishnav (or devotee of God Vishnu) who was also greatly influenced by Jainism, Gandhi was born into an affluent family in the western part of Gujarat and educated as a lawyer in England. He opposed the injustices of colonial rule, and his greatest contribution toward its end was his critique of the industrial and consumer-oriented society. In his book, *Hind Swaraj* (2009 [1909]), Gandhi criticized civilization that was based on the promotion of material things, writing, 'This civilization is irreligious (*adharma*) and it has taken such a hold in the people in Europe that those who are in it appear to be half mad' (p. 34). Gandhi was greatly influenced by Tolstoy, Ruskin, and Thoreau and his own reading of the *Bhagavad Gita*. According to Kent (2016), 'Gandhi argues that true civilization entails not the speedy satisfaction of individual desires, but the restraining of desire through a sense of duty and humility' (p. 301).

As a social reformer, Gandhi was influenced by many different sources, and his personal philosophy is rooted in his religious faith. In his reading of the *Gita*, he held the view that 'all embodied life is in reality an incarnation of God' (Gandhi, 2000 [1926], p. 17; Kent, 2016, p. 301). For Gandhi, the way to self-realization is to recognize the divinity in others and to love them as oneself. His political activism of working for disempowered people in society came from this realization. In addition, he realized that the fulfillment of material desires only is selfish, putting this realization into practice in his own life.

Gandhi's strong belief was that the Hindu dharmic civilization, which is based on codes of right behaviour, can be restored by living a simple life in which the luxury of consumer oriented, mass-produced, machine-made goods is not pursued. He was in favour of weaving the cotton out of which his clothes were made, using this to boycott British-made products and to support his strategy of nonviolent non-cooperation (Kent, 2016).

1.2.1 Nonviolence and Gandhi's Truth

The major difference between Gandhian and non-Gandhian systems of nonviolence is a disagreement about the relation of ends to means. To Gandhi, the selection of means precedes the selection of ends in the sense that one's original and basic commitment is to certain means or methods, and one's ends are seen as objectives that emerge progressively as one takes the opportunity to understand an opponent and one's own mission. In the Gandhian scheme of *satyagraha*, one undergoes self-suffering in the belief that an opponent can be converted to seeing the truth by touching his or her conscience or that a clearer vision of truth may grow out of the dialectical process for both parties. *Tapasya* or self-discipline is necessary to achieve the highest truth. It involves freedom from fear and a willingness to die. The suffering that has to be undergone in *satyagraha* is *tapasya* in its fullest form.

Tapasya was on full display on 12 March 1930, when Gandhi started the well-known Dandi march as a way to protest against the salt law imposed by the British. Thousands of people, including women and children, joined the march, ultimately subjecting hundreds of people to a brutal British response. Gandhi had advised the people to make salt by themselves in their own homes as a protest against the government. On 6 April 1930, he broke the British salt law by picking up a chunk of the salt crust from the Dandi beach. It was a unique method of protest and taught the world that the power of nonviolence supersedes the power of violence.

Gandhi's concept of *satyagraha* was embraced by the environmental movements in India, making Gandhi an environmental icon. Yet in his writings of the 1920s, 1930s, and 1940s, we find ample evidence that Gandhi's primary concern was social reform and political mobilization. As Guha (1995) observes, 'his reservations about the wholesale industrialization of India are usually ascribed to moral grounds – namely, the selfishness and competitiveness of modern society – but they also had markedly ecological undertones' (p. 49). Consider the following remarkable passage from Gandhi's journal, *Young India*, dated 20 December 1928:

> God forbid that India should ever take to industrialization after the manner of the West. The economic imperialism of a single tiny island

> kingdom (England) is today keeping the world in chains. If an entire nation of 300 million took to similar economic exploitation, it would strip the world bare like locusts. https://www.gandhiheritageportal.org

Two years earlier, Gandhi had claimed that to 'make India like England and America is to find some other races and places of the earth for exploitation'. As it appeared that the Western nations had already 'divided all the known races outside Europe for exploitation and there are no new worlds to discover', he pointedly asked, 'What can be the fate of India trying to ape the West'? (Gandhi, *Young India*, 7 October 1926) The answer to his question is by now painfully obvious, for in the past few decades, we have attempted precisely to 'make India like England and America'. Without the access to resources and markets enjoyed by those two nations when they began to industrialize, India has had to rely inevitably on the exploitation of its own peoples and environment. The natural resources of the countryside have increasingly been used to meet the needs of the urban-industrial sector, and this diversion of forest, water, and other resources to the elite has accelerated the process of environmental degradation while at the same time depriving rural and tribal communities of their traditional rights of access and use. Meanwhile, the modern sector has moved aggressively into the remaining resource frontiers of India, the northeastern states and the Andaman and Nicobar Islands (Guha, 1995).

1.2.2 How Chipko Followed Hinduism through Gandhi

With his leadership of the Chipko movement, Chandi Prasad Bhatt demonstrated Gandhian values by advocating the conservationist environmental view that the Himalayan forests should be used in a sustainable way so that it benefits the local people and not the big corporations that destroy the forest. Bhatt was a follower of Vinoba Bhabe (1895-1982), who is sometimes called the spiritual successor of Gandhi's *sarvodaya* (well-being of all) movement and is well-known for his *bhoodan* (land gift) movement as part of *sarvodaya*. By following *sarvodaya*, Bhatt became familiar with the suffering of the hill peasants in the upper Himalayas, where the Chipko movement began. Kent (2016) notes that 'His experience in social organizing helped him to channel local people's long-simmering resentments into effective challenges to state's extractive economic policies' (p. 302).

Bhatt has continued his efforts with the non-governmental organization Dashuli Village Self-Help Association (Dashuli Swarajya Seva Mandal), which advocates economic self-reliance, women's rights, and developing technology to improve the lives of the people in his local community. Kent points out that Bhatt follows the values held not only by Gandhi but also by other social reformers like Sri Aurobindo and Swami Vivekananda, who have given a more practical meaning to religion with the viewpoint that by serving others, one serves God. This has resonated with the general Hindu population.

Sunderlal Bahuguna, another Chipko leader, promotes the preservationist view of the environment, believing that the wilderness should not be interfered with. He follows the prophetic-ascetic side of Gandhian activism, arguing for the inherent and sacred value of nature (Guha, 2000). According to Kent, 'With his long beard, white garments, and values of simple living, Bahuguna invokes a familiar Brahmanical Hindu paradigm of the renouncer-guru whose distance from hurly-burly of profane life endows him with the authority to critique society in a disinterested and thus more truthful way' (2016, p. 303). Bahuguna used the Gandhian non-violent strategies of foot marches, mass demonstrations, and fasts and conducted 'readings of *Bhagavad Gita* to instill in the participants and convey to outside observers the righteousness and moral superiority of their cause' (Kent, 2016, p. 303). Kent notes that while Bahuguna's protest against the Tehri Dam was not successful, some claim that his action resonated with the Hindu extremist groups like the Vishwa Hindu Parishad (VHP) and Rashtriya Swayamsevak Sangh (RSS) due to his use of the language of religion (see Sharma, 2014). Kent (2016) cautions against unexpected side effects of following religion.

1.2.3 How NBA Followed Hinduism through Gandhi

According to Fisher (2000), 'In the Narmada controversy we find every reference to Gandhi except, perhaps, his devotion to truth...' (p. 416). The residents of Narmada valley consider the Narmada river as sacred and treat it with respect and reverence. The thousands of temples along Narmada are considered a pilgrimage. In the Narmada struggle, this sacred view of nature is set in opposition to a view of nature as a resource that needs to be used for economic growth and as a technological solution to the issue of providing water and electricity to the drought-stricken areas of Gujarat. This latter view 'reflects Descartes' conviction that the general good of all humankind could be pursued by the attainment of knowledge that is useful in life so as to make ourselves "the masters and possessors of nature"' (Fisher, 2000, pp. 409-410). Fisher argues that, first, the Narmada controversy has been oversimplified as a political debate between the developers and the resisters and resulting in transforming the resources from those who use the resources in a sustainable way to those who use the resources in an unsustainable way to benefit themselves at the cost of the environment. Second, he says that the transformation of resources have been done in such a way that it leads to the mastery and reshaping of nature. Third, 'the process allows and even requires that governments consolidate their control over both resources and people' (p. 411). Fisher cautions that though the rhetoric of Gandhi can open up possibilities but it can also direct attention away from the conflict.

1.2.4 How Navdanya Followed Hinduism through Gandhi

The Earth Democracy that Navdanya promotes asks people to make a paradigm shift from the traditional approach of the human conquest of nature to one of living in harmony with it, living on Earth as a member of one family, and having healthy eating habits to improve the well-being of this planet in all its aspects. Founder Vandana Shiva (2017) says, 'Organic agriculture is the *Dharma* ("path of righteousness") that sows the seeds of peace and prosperity for all. It helps us break out of the vicious cycle of violence and degeneration, and create virtuous cycles based on nonviolence and regeneration' (p. 18). Pradeep Dhakal (2015) adds, 'The Hindu concept of *Vasudhaib Kutumbakam* is essential in this regard. This concept does not define or proclaim any religion. It is a humanistic approach for world peace and human welfare' (p. 160).

1.2.5 Ecology and Social Justice

People's environmental movements in India, without doubt, are concerned with the continual deterioration of resources and the need to protect them, but they have also insisted on the moral rights of communities with respect to natural resources and their equitable distribution and sustainable use. Concurring with Sangvai's (2007), it can be argued that the environmental movements 'have added distributive justice concerns and principles with respect to the right over resources – land, forests, and water bodies – as a major criterion for socio-economic equality' (p. 113). They are not only questioning the unequal distribution of resources but also about the technology of production, by highlighting that it is the bottom-up decision-making and empowerment that should be promoted. The power-dynamics need to be changed from centralization to decentralization.

It must be stressed that Indian environmental movements are fundamentally different from those in the West because they are triggered by the local poor's struggle for survival. As Misra (2007) says,

> Environmental movements in India are therefore not exclusively concerned with 'green' and 'clean' earth, or with saving endangered species and preserving the 'land ethic' as in the West. Such movements are motivated above all by the concern for the very survival of the local poor. (p. 138)

The Chipko movement, discussed in Chapter 2, and Narmada Bachao Andolan, discussed in Chapter 3, connect social justice and women's rights with environmental concerns, allowing the needs of the poor to be addressed alongside environmental issues. By doing so, they have been successful in uniting Hindu religious values and tribal culture (Misra, 2007).

The problems related to the use and conservation of natural resources in developing countries such as India are of a different nature than those of developed countries. In developed countries, the primary issue is protection of what remains in nature; in India, the preservation of natural resources must necessarily consider the competing claims of humans on these resources for their sustenance and livelihood. And this includes a huge population that is the poorest in the nation and is completely dependent on the forests as its life-support system. In recognition of the genuine relationship between the extant forest resources and the human beings who depend on them, any legal administrative regime must aim to judiciously utilize these resources to ensure sustainability (Hazra, 2002).

Indian environmental movements have perceived the issues of social justice and equity as their central goals. There is a tendency to label any social movement that grows out of a conflict over the use of natural resources between state and industry on the one hand and local people on the other as ecological. It can be argued that social movements provide powerful and eloquent evidence that the voices of depressed, disadvantaged, and disenfranchised groups of people ought to matter and that the individuals in such groups aspire to be recognized as agents fully responsible for their future and actively trying to change their situations rather than as victims. Secondly, those who are aware of the destruction of nature and the disappearance of indigenous cultural traits in the name of economic development consider the ecological movements as a way out of their dilemma; such movements offer a glimmer of hope (Linkenbach, 1994). In short, it is not so much a question of which parts of non-human nature ought to be protected and preserved;

it is rather the case that issues of maldistribution and overconsumption of resources dominate the green agendas of Indian environmentalism.

In developing countries such as India, environmentalism has acquired a new meaning. Conservation ecologists compile information as they do in the West, inculcating sensitivity regarding flora and fauna through the study and appreciation of birds, trees, animals, mountains, rivers, the sea, and the entire globe and its atmosphere. For the most part, such environmentalism is devoid of any socio-political linkages. There is also a managerial environmentalism among the political and industrial-urban elite for whom environmental problems are considered issues to be managed and alleviated by means of techno-economic intervention without requiring changes in the basic framework of the system.

Comparing environmental movements in the United States and India, Chapple (1998) observes that '[w]hereas in the American context, the early rallying cry for environmental action came from scientists and social activists with theologians only taking interest in this issue of late, in India, from the outset, there has been an appeal to traditional religious sensibilities in support of environmental issues' (p. 20). Moreover, environmental movements in India have been nonviolent in character and firmly embedded in existing social and political structures. Unlike their Western environmental activist counterparts, Indian movements are also strongly tied to human rights, women's rights, social justice, and equity, which gives them a distinct character. Undoubtedly, this distinct character of Indian environmentalism is linked to the Indian ecological vision of life (Misra, 2007).

1.3 Conclusion

This chapter has presented an overview of multiple Forest Acts implemented by the British Raj still enforced today, showing how the people living around the forests were deprived of their rights to the forest by the British rulers in the name of shipbuilding and building railroads, bridges, etc. Since independence, some attempts have been made to give back some of these rights to the people living around the forest, but the implementation of such attempts has been challenging even to this day. An analysis of how the three environmental movements – Chipko, the NBA, and Navdanya – have followed Hinduism through Gandhi have been explored. The chapter concludes with a brief explanation of how environmentalism in India is different from Western environmentalism. The primary difference is that Indian environmentalism is to a large extent related to the survival needs of the poor indigenous tribal people and is focused on demanding that the rights of the marginalized be taken into consideration, whereas environmentalism in the West is mostly related to the protection of nature.

We turn next to the renowned Chipko movement, in which indigenous women who depended on the forest for fodder and sustenance took their leaders' advice to hug the trees. This simple action sparked a movement that resulted in a fifteen-year ban on tree-cutting in the Uttarakhand region of the Himalaya.

Works Cited

Chapple, Christopher Key (1998). 'Towards an Indigenous Indian Environmentalism'. In Lance E. Nelson (ed.), *Purifying the Earthly Body of God.* Albany, NY: State University of New York Press, 13-38.

Descartes, Rene (1971). *Discourse on Method and the Meditations,* translated by F.E. Sutcliffe. Hammondsworth: Penguin.

Dhakal, Pradeep (2015). 'Hindu Vision of Nonkilling: Realizing Truth for Creating a Nonkilling World'. In Joam Evans Pim et al. (eds.), *Nonkilling Spiritual Traditions,* First Edition. Honolulu, HI: Center for Global Nonkilling, 151-160.

Fisher, William F. (2000) 'Sacred Rivers, Sacred Dams: Competing Visions of Social Justice and Sustainable Development along the Narmada'. In Christopher Key Chapple and Mary Evelyn Tucker (eds.), *Hinduism and Ecology: The Intersection of Earth, Sky, and Water.* Cambridge, MA: Harvard University Press.

Fernandes, Walter, and Vijay Paranjpya (1997). 'Hundred Years of Involuntary Displacement in India: Is the Rehabilitation Policy an Adequate Response'. In Walter Fernandes and Vijay Paranjpya (eds.), *Rehabilitation Policy & Law in India: A Right to Livelihood.* New Delhi: India Social Institute, 1-34.

Gadgil, Madhav, and Ramachandra Guha (2003). *This Fissured Land: An Ecological History of India.* New Delhi: Oxford University Press.

Gandhi, Mohandas Karamchand (1926). *Young India,* October 7, Gandhi Heritage Portal, https://www.gandhiheritageportal.org/.

— (1928). *Young India,* December 20, Gandhi Heritage Portal. *https://www.gandhiheritageportal.org/*

— (2000) [1926]. *The Bhagavad Gita According to Gandhi.* In John Strohmeier (ed.), Berkeley, CA: Berkeley Hills Books.

— (2009) [1909]. *Hind Swaraj & Other Writings.* In Anthony Parel (ed.), Cambridge: Cambridge University Press.

Guha, Ramachandra, and Madhav Gadgil (1989). 'State Forestry and Social Conflicts in British India'. *Past & Present,* 123, 141-177.

Guha, Ramachandra (1995). 'Mahatma Gandhi and the Environmental Movement of India'. *Capitalism, Nature, Socialism*, 6, 3, 47-61.

— (1999). *Environmentalism: A Global History.* New York: Longman Publishing Group.

— (2000). *The Unquiet Woods: Ecological Change and Peasant Resistance in the Himalaya.* Berkeley, CA: University of California Press.

— (2001). 'The Prehistory of Community Forestry in India'. *Environmental History*, 6, 213-238.

Hazra, Arun Kumar (2002). 'History of Conflict over Forests in India: A Market Based Resolution'. *Working Paper Series, Julian L. Simon Centre for Policy Research*, 1-46.

Kent, Eliza (2016). 'Hinduism and Environmentalism in Modern India'. In Brian A. Hatcher (ed.), *Hinduism in the Modern World.* New York: Routledge, 290-308.

Leopold, Aldo (1966). *A Sand County Almanac.* Oxford: Oxford University Press.

Linkenbach, Antje (1994). 'Ecological Movements and the Critique of Development: Agents and Interpreters'. *Thesis Eleven*, 39, 63-85.

Misra, Shalini (2007) 'Spirituality, Culture and the Politics of Environmentalism in India'. *The Journal of Entrepreneurship*, 16, 2, 131-145.

Nayak, Arun Kumar (2015). 'Environmental Movements in India'. *Journal of Developing Societies*, 31, 249-280.

Sangvai, Sanjay (2007). 'The New People's Movements in India'. *Economic & Political Weekly*, 42, 111-117.

Saxena, K.B. (2008). 'Development, Displacement & Resistance: The Law & Policy on Land Acquisition'. *Social Change*, 38, 3.

Sharma, Mukul (2014). *Green and Saffron: Hindu Nationalism and Environmental Politics.* Delhi: Orient Black Swan.

Shiva, Vandana (2017). 'Swaraj: From Chipko to Navdanya'. In John Hart (ed.), First Edition. *The Wiley Blackwell Companion to Religion and Ecology.* John Wiley and Sons Ltd., 12-19.

2 Chipko (Hug the Trees) Movement

Abstract
This chapter examines the iconic Chipko movement of the 1970s that protested against deforestation in the Himalayan region using the nonviolent gesture of hugging the trees, thereby showing the interconnectedness of the poor women's livelihood with natural resources. The movement had three strands: one led by Chandi Prasad Bhatt, the second led by Sunderlal Bahuguna which applied the Gandhian method of *satyagraha* and was embraced mostly by poor women of the region, and the third being the Marxist and radical strand which was later dissolved. The chapter explores how the Chipko movement emerged initially as a peasants' movement but later became quite complex and attracted great public support within India and outside. This was the first time that the environmental aspect was introduced into the development discourse.

Keywords: Chandi Prasad Bhatt, Sunderlal Bahuguna, Gaura Devi, social justice, the myth of Chipko

Although the Chipko movement holds a special place in environmental history as the first demonstration of ecofeminist ideas in action, the movement evolved from a long history of challenges to the demands of commercial forestry. By 1921, as Guha (1989) reports, there was 'a near total rupture between the colonial state and its subject population' (p. 137). Peasant protests against the Forest Acts and government attempts to break them up or calm them down continued for decades. The activists' protest tactics included a variety of nonviolent models such as strikes, 'go-slow' protests by porters for visiting officials, refusals to pay fines, and subversion (the standard technique of burning litter on the forest floor to open it up for grazing had been banned in order to improve tree regeneration) (Guha, 1989). The protests were rebellious and met with state violence, notably in Tehri Garhwal in the 1940s.

Mallick, Krishna, *Environmental Movements of India: Chipko, Narmada Bachao Andolan, Navdanya*. Taylor & Francis Group, 2021.
DOI: 10.5117/9789462984431_CH02

Decades later, continued conflict over access to forests and the allocation of felling rights as well as ecological problems like landslides and floods led to a series of nonviolent actions to prevent contractors from logging forests. Scholars delineate three streams within the broader Chipko movement: one in the Chamoli area led by Chandi Prasad Bhatt, who was deeply influenced by socialist ideology; another led by Sunderlal Bahuguna in Garhwal; and a third led by Uttarakhand Sangharsh Vahini (USV) in Kumaon. Being more Marxist in its orientation, the third strand provided a radical focus to the movement.

2.1 A Physical Act of Survival

The protests against deforestation in the Himalayan region of India ultimately showed how women's livelihood are intertwined with natural resources. Among these, it was the women's initiative to hug the trees in the village of Reni in 1974 that became the centrepiece of a global myth. While this myth glamorized Chipko as an environmental and women's movement, in actuality, Chipko was merely the concerted reaction of peasants who were concerned about their survival (Guha, 1989). Still, the action clearly represented a naturalist worldview; Guha writes in *The Unquiet Woods* that 'peasant movements like Chipko are not merely a defense of a little community and its values, but also an affirmation of a way of life more harmoniously adjusted with natural processes' (p. 196).

In 1970, resentment among the people of Uttarakhand regarding deforestation came into clear focus, expressed by the formation of a powerful and influential social movement. A catastrophic flood in the Alakananda (the sister of the river Ganges in Indian mythology) valley convinced the villagers of the devastating ecological consequences of the monumental deforestation. That same year, the Forest Department authorized the commercial felling of 2,000 trees in the Alakananda valley and refused to allocate ash trees to the villagers for household purposes. It was at this point that Chandi Prasad Bhatt, one of the leading activists of the Chipko movement, recommended using what has now become the historic Chipko protest strategy of hugging the trees in order to protect them. Due to the protest, commercial felling in the upper drainage area of the Alakananda river and its tributaries was banned for a period of ten years (James, 2000; Misra, 2007). Only as Chipko grew did the movement show more interest in ecological issues such as the depletion of forest cover and soil erosion. And it was only after it drew the attention of international scholars that it came to be regarded as an environmental movement.

2.1.1 Sacred Texts and Social Justice

Chipko is unique in its spiritual and moral dimensions. Misra (2007) mentions that the protests were accompanied by readings of the *Bhagavad Gita* and other ancient Indian texts that attest to the role of forests in Indian life as well as the recitation of poetry and folk songs acclaiming the value of forests and human beings' place within them. The tying of *rakhees* (wrist bands generally used during Indian festivals) by women around trees represented a bond of kinship with them (James, 2000; Misra, 2007). According to Misra (2007), the Chipko movement was, in essence, a peasant uprising that used traditional methods of nonviolent resistance as a form of protest against the loss of a vital source of livelihood.

Insofar as its members had enjoyed a longstanding symbiotic relationship with the natural surroundings, the Chipko movement was not an environmental movement in the familiar Western sense of the term. Rather, the movement was ultimately the people's reaction to the destruction of their habitats, which was a threat to their very survival. It is morally wrong in the Rawlsian sense to take away people's livelihood. And this is certainly the case in the Himalayan region, where the people have been exploited and made worse off by actions taken by both the central and local government who have continued with the colonial-era forest policies.

An analysis of the different aspects of how this movement was shaped is important to understand the complexities of the movement (Misra, 2007). The formidable use of religious symbols, Gandhian economic and political ideologies, and methods of nonviolent resistance were embedded in the Chipko movement. Another positive aspect was the active participation of all social groups. The Bajgis, a group of musicians, were solicited to mobilize the villagers through their drums. Although many of the movement's leaders were men, women also played a very prominent role because they were the ones most strongly affected by the rampant devastation that led to the shortage of fodder, firewood, and water for drinking and irrigation. Even children joined the movement.

The Chipko movement was politically quite diverse. It was also remarkably effective, achieving a fifteen-year ban on the felling of green trees above 1,000 meters (3,280.84 feet) altitude in 1981. However, it is the wider ramifications of the Chipko movement – both within India and globally – and, more particularly, the beliefs of the movement that have shaped its reception in the national and international arenas. Bandyopadhyay (1992) describes the way in which Chipko metamorphosed from a peasant movement to a global campaign for the sustainable management of forests in general and those of the Himalayas in particular. He suggests that Chipko is 'no longer a hill people's movement against forest fellings' (p. 300) but that it has become a philosophy and, more precisely, an extension of Gandhian thought. As Shiva and Bandyopadhyay (1986) have noted, 'The key agenda of the Chipko movement was to carry forward the vision of Gandhi's mobilization for a new society, where neither man nor nature is exploited and destroyed, which was the civilizational response to a threat to human survival' (p. 21). The main demand of participants was to have access to the benefits of the forest, especially the right to fodder (Bahuguna, 1990).

As a local movement in pursuit of livelihood protection, the Chipko protests had earlier antecedents. However, it is on account of its association with nonviolent Gandhian techniques and the sacred geography of the Himalayas and the Ganges that the movement acquired a wider and more formidable symbolic significance. The Chipko movement was inspiring to social and political activists, giving rise to many other movements including Appiko, Narmada Bachao Andolan, Save Western Ghats, and Jungle Jeevan Bachao. Numerous government development plans were the trigger for these organized grassroots protests.

2.2 Preconditions and Formation of the Chipko Movement

The Uttarakhand Himalaya, where the Chipko movement had its genesis, has a history of environmentally friendly and economically sustainable development. Formerly a part of the State of Uttar Pradesh, it is a hilly, mountainous area with very steep slopes that used to be covered with dense forests of predominantly broad-leaved trees. Two or three harvests per year had been possible throughout the course of the last century, with wheat, rice, and millets being among the chief grains grown by the local population. The systems of the village and the method of crop rotation constituted a major part of the hill people's natural environment and way of life. The production of crops was geared towards the satisfaction and maintenance of the most basic subsistence needs, and for many generations, these needs were comfortably met. There also remained a surplus of grain for export. The diet of the people of Uttarakhand was supplemented by fish, fruit, vegetables, and animal flesh (Guha, 1989).

The way of life that the Uttarakhandi people cultivated over an extensive period of time suited a very delicate ecosystem. Hilltops were dedicated to local deities, and the forest around the designated places of worship was regarded as sacred. Although many of the wooded areas were the outcome of spontaneous natural growth, those areas also displayed evidence of the hill people's 'instinct for the plantation and preservation of the forest' (Guha, 1989, p. 29). The people there enjoyed almost complete control over their forest. They showed a deep love for vegetation and felt a sense of responsibility with respect to future generations. The community was homogenous and had a democratic form of self-government. The community's attitude toward the environment was profoundly shaped by the belief that one ought to live sustainably and take proper care of the available natural resources.

European capitalism arrived in India in the form of colonialism under the British Raj in 1858. The British exhibited very little concern for the local people and their rights. Relying on the supposed 'right of conquest' as a justificatory principle, they claimed ownership of the forests that in fact belonged to the people of Uttarakhand. The devastation that followed was significant. The heavily forested Garhwal and Kumaon hill areas endured the brunt of British scientific forestry practices that replaced the natural mixed forests – which had long supplied the local population with fuel, fodder, and fertilizer – with exotic monocultural plantations growing a single crop. The introduction of commercial forestry led to deforestation, the gradual destruction of traditional subsistence agriculture, and, eventually, the large-scale emigration of the people of Uttarakhand. The British colonial government had thus not only violated indigenous people's rights to their forest, it had also created a growth-based economy that contributed to the region's devastation and destabilization.

The adoption of scientific forestry in both the traditionally governed Tehri Garhwal and the colonially governed Kumaon as well as the reservation of forests for timber production towards the end of the 19th century led to the considerable disruption of production systems and hardship for the peasants in this area (Guha, 1989). In both regions, peasants protested in different ways and with different effects against their exclusion from forest ownership, and these protests continued throughout the course of the 20th century. Post-independence India continued commercial forest exploitation with even greater vigour than during the colonial era. The reach of the Forest Department and its contractors spread to the remotest corners of the country with the expansion of the road network. Local livelihood received even less attention than they did under colonial rule because state policy consistently favoured the export of raw timber and resin for processing by large industry in the plains.

By the 1970s, the Chipko movement had emerged to demand that priority be given to local employment in the extraction and processing of forest produce (Guha, 1989). Sarin notes that the increasing incidents of landslides and floods and the declining availability of biomass for the satisfaction of basic subsistence needs motivated even hill women to join the movement, thus broadening the popular base of Chipko protests. 'Ironically, the Kumaon *van panchayat* (community forestry management) rules of 1931 were revised in 1976, at the height of the Chipko movement, substantially reducing *van panchayat* authority and entitlements even over village forests' (Sarin, 2001, p. 346). The issue of local forest rights, however, was soon superseded by the new national and global ideology of environmental conservation. Instead of giving priority to local forest-based livelihood, the Forest Conservation Act of 1980 empowered the central government to make decisions related to even the smallest patch of forest land, under community forestry management's control. The Uttar Pradesh Resin and Forest Produce Act of 1975 made the tapping, sale, and purchase of all resin a state monopoly (Sarin, 2001).

2.2.1 Workers Organize for Nonviolent Action

A decade earlier, a workers' cooperative of unskilled and semi-skilled construction workers had been founded by Bhatt and his co-workers in the Chamoli district of Uttarakhand. The cooperative had functioned successfully for some time but was eventually outdone by rich contractors who managed to become strong lobbyists for themselves in official circles. In 1964, venturing to generate more employment opportunities by opening small workshops designed to make farm tools for local use, the workers' cooperative established the Dasholi Gram Swarajya Sangha (DGSS) organization – a local cooperative based in the Chamoli districts – and entered the market by buying rights over forests through auction. Yet here again, after initial success, the group was overtaken by other contractors.

Chandi Prasad Bhatt and his co-workers belonged to the Chamoli district and cared primarily about increasing employment avenues for the local people. They embraced the ideology of nonviolence, *satyagraha,* as advocated by Mahatma Gandhi. Bhatt and his followers also called their initiative the *Sarvodaya* movement, after Vinoba Bhave, a later Gandhian activist, who called his social movement of post-independent India, the *Sarvodaya* (welfare for all). *Sarvodaya* was a key concept in Gandhi's political philosophy and *Sarvodaya* was the title of Gandhi's translation in Gujarati (Gandhi's native language) of John Ruskin's *Unto the Last.* Then, in early 1973, the Forest Department refused to allot ash trees to the DGSS for the making of agricultural implements, giving this concession instead to the private Simon Company. This triggered a powerful reaction by DGSS members who decided to fight against this perceived injustice by lying down in front of timber trucks and burning resin and timber depots – actions that were similar to those taken in the Quit India movement. When the employed methods proved ineffective, Bhatt suggested embracing the trees, thus giving the movement its name – the term 'Chipko' means literally 'to hug' (Bahuguna, 1990; Guha, 1989). This form of protest was instrumental in driving away Simon Company from involvement in the felling of the ash trees. After its local success, the movement spread to other neighbouring areas and became, in the long run, internationally renowned.

During the initial phases of the movement's advance, the government's response was not encouraging. In January 1974, the government decided to auction over 200 trees in the Reni Forest situated in the Alakananda valley. The valley had been severely affected by landslides in the recent past. The villagers began immediate protests against the government's auction by hugging the trees. Over the next few weeks, rallies and meetings continued in the Reni area. The landmark event in this struggle took place on 26 March 1974, when a group of peasant women in the Reni village of Uttarakhand acted to prevent the cutting of trees and, in doing so, defended their traditional forest rights that the contractor system of the state forest officials had threatened and undermined. Although forest officials in Reni resorted to dishonest actions, the villagers were nonetheless able to save the forest. Full non-cooperation tactics were adopted in the entire movement, and no question of using violence ever arose. Rallies and meetings were continuously held, and a reading of the *Bhagavad Gita* was also organized. Along with these strategies, women tied ornamental wrist bands used during festivals (known as *rakhees*) around the wounded spots on the trees, thus signifying their close relationship with the trees.

2.2.2 Suffering by Means of Fasting and Foot March

In 1974, under the leadership of Sunderlal Bahuguna, more than 3,000 women and men of the Tehri Garhwal district – situated on the foothills of the Himalayas – organized a major demonstration against the extreme drawing of sap from chir pine trees by cutting into them (Gosling, 2001; James, 2000; Misra, 2007). Bahuguna, a *sarvodaya* or 'welfare for all' leader, used Gandhian principle of nonviolence, *satyagraha,* with its methods, such as fasting and a long and vigorous foot march (*padayatra*) along the full length of the Himalayas, from Kashmir in the northwest to Kohima in Nagaland in the northeast (Misra, 2007). The aim was to encourage people to understand the significance of protecting their forests. Similar to Gandhi, Bahuguna was capable of addressing secular issues such as the importance of environmental conservation from within a spiritual framework shaped by Hindu scriptures and inspired by the religious principles of the Buddha and Vivekananda, on the one hand, and by the work of distinguished Indian scientists such as Jagdish Chandra Bose, on the other (Gosling, 2001). Misra (2007) argues that Bahuguna's personal philosophy was a unique characteristic of the moral and religious dimension of the Chipko movement.

Despite the success of previous Chipko agitations, in 1979, a private contractor was awarded a big contract for the felling of trees in the Badyagarh Patti of Tehri Garhwal district. The felling was very destructive, and it happened primarily because of the erosion of the local community's spirit of cooperation and solidarity. At this juncture, Sunderlal Bahuguna – along with the *Sarvodaya* workers Dhum Sing Negi, Kunwar Prasun, Pratap Sikhar, and Vijay Jardari – came to Badyagarh and began rallying the people by pointing out the harmful ecological consequences of the felling. At the same time, Bahuguna's wife Vimla and other women mobilized the village women on the issue. The campaign reached a new height when Bahuguna began to fast on 9 January 1979, instilling tremendous inspiration among the villagers. Bahuguna was arrested on 22 January of the same year, but even after the removal of their leader from the scene, the determined resistance of the villagers forced the contractor and the forest officials to admit defeat and abandon the felling.

After the Reni incident, the Chief Minister of Uttar Pradesh, H.N. Bahuguna (no relation to Sunderlal Bahuguna), agreed to set up a committee headed by a botanist, Virendra Kumar, to investigate the matter. A second government committee headed by K.M. Tewari was also formed. It was in the glow of the Chipko movement that Sunderlal Bahuguna became an internationally recognized environmental activist travelling with the Chipko message to Europe, Africa, and the United States. But despite the detailed project reports prepared by both committees, their recommendations were never implemented by the government. However, in April 1981, when Sunderlal Bahuguna went on an indefinite fast demanding a total ban on the green felling in the Himalayan forests of Uttarakhand, the area was finally declared a protected realm. This particular victory was the greatest achievement of the movement.

2.3 Laudable Leaders

International ecologists saw the Chipko movement as a cultural response of the Indian indigenous population as well as one inspired by profound affection for the environment. When the Swedish parliament presented the Right to Livelihood Award to Chipko activists on 9 December 1987, the text of the award read:

> The Chipko Movement of India, whose members 'hug trees' to prevent their felling, and have revived traditional agro-forestry, is honoured for its dedication to the conservation, restoration and ecologically-sound use of India's natural resources. (Partial text of the award can be found at https://www.rightlivelihoodaward.org/laureates/the-chipko-movement)

The movement's leaders were also awarded by the Indian government. Bahuguna received the prestigious Padma Vibhushan award in 2009, the second largest civilian honour, and Bhatt was awarded the Padma Bhushan award in 2005, the third highest civilian award. Bhatt had received the Magsaysay Award for Community Leadership earlier in 1982. Of course, it is misleading to think of Chipko itself as only a 'Gandhian' movement insofar as its roots can be traced to a century-old tradition of peasant resistance in defense of their forest rights. However, Bhatt and Bahuguna – the movement's best-known leaders – exemplify the insight of Gandhi's constructive work.

Urban admirers of Chipko are usually identified as being supporters of either Chandi Prasad Bhatt or Sunderlal Bahuguna, but there is in fact ample reason to celebrate both men. Bhatt and his organization – the Dashuli Gram Swarajya Sangha (DGSS) – played a seminal role in the origins of Chipko. It was Bhatt himself who suggested the tree-hugging technique of protest to the peasants of the upper Alakananda valley. Since the early days of coordinating protests against commercial forestry, the DGSS has focused increasingly on environmental restoration. The organization has also taken a leading role in uniting women for afforestation work in the villages of the Alakananda valley, where the organization's tree-planting and protection programmes have been a great deal more successful than the lavishly funded schemes of the state Forest Department.

Guha and Martinez-Alier (1997) argue that Chandi Prasad Bhatt must be considered the pioneer of the Chipko movement. But Sunderlal Bahuguna has a long record of social work that goes back even prior to the Chipko movement. Bahuguna and his wife Vimla were among the *Sarvodaya* workers trained by Sarla Devi (Catherine Mary Heilman), a remarkable English disciple of Mahatma Gandhi who moved to the hills in the 1940s. From his home base in the Bhageerathi valley, Bahuguna organized several important Chipko protests between 1977 and 1980, and in the years since, Bhatt and his colleagues have concentrated on ecological restoration within the hills.

Bahuguna has travelled widely in India and abroad. 'He is a captivating speaker too, and in this capacity has done a great deal to alert the urban intelligentsia to the dangers of unbridled materialism and consumerism (Guha et al., 1997, 154).

According to Guha and Martinez-Alier (1997), the example of Gandhi animates the lives of both Bhatt and Bahuguna, but each of them has taken something distinctive from the life of the master: 'Bahuguna is a prophet and moralist who appeals, as Gandhi did, to the conscience of individuals, urging them to abjure consumerism and return to a simpler way of life' (p. 154). Using a different approach by working out in practice a sustainable economic alternative to centralized development, Bhatt and his group of followers have emphasized Gandhi's constructive work of training activists in Gandhi's own Sabarmati and Wardha ashrams. Chandi Prasad's work has helped infuse a new ecological meaning into the Mahatma's ideal of *gram swaraj* (village self-reliance).

The Chipko movement was particularly successful, Rangan claims, when its leaders articulated the movement's demands in the state's vocabulary of national integrity, development, and democracy and when they combined this language with symbolic acts of popular protests. The protests gained wider audiences through simple, popular narratives that pitted peasants against the state and markets while at the same time glossing over the heterogeneity of classes, interests, and constituencies within the movement itself. As Rangan (1996) posits, 'This skilled interweaving of state and populist rhetoric made Chipko the unquestioned icon of grassroots environmentalism in India and a model for international environmental circles' (p. 216). Environmental scholars reframed the Chipko story as India's cultural response to the ecological crisis in the Himalayas. International NGOs and scholars praised the Chipko movement as an inspiration for environmental activists around the globe, claiming that the ideals it represented were far more important than the aims it initially had set out to achieve. The movement assumed legendary status when, following appeals to the Indian government and drawing widespread attention from scholars and the media, several pieces of legislation and constitutional amendments aimed at forest protection were introduced between 1975 and 1980 (Rangan, 1996).

2.3.1 Women's Role in the Chipko Movement

When the Chipko movement spread to other regions of India and eventually the world, scholars began to view it as an environmental movement embraced by women. Some critics have called this an instance of neo-populist theorizing, and others have called it the 'myth of Chipko'. The poor women of the Himalayan region were yearning for freedom from ecological oppression. It can be pointed out that by using the forest in a sustainable way to fulfill their survival needs, the women of Chipko showed that fulfilling these needs and sustaining the forest's biodiversity are not mutually exclusive endeavours. They also demonstrated that they must have a role in the making of forest policies, as they are the ones directly impacted by them, not elites in power who are not directly affected by deforestation.

Jain (1984) points out that one reason why it was primarily women rather than men who saw the strong connection between the disruption of that balance and the possibility of survival has to do with the way the local subsistence economy was organized. Burdened with the responsibility of crop cultivation, livestock, and childcare, women were most directly affected by the environmental disasters caused by deforestation in the hill areas of Garhwal. The message of the Chipko workers appealed to these women because they understood the link between their victimization and the devastation wrought on the mountain slopes by commercial interests. Women clearly saw the reasons behind their existential problems. The need for sheer survival compelled them to support a movement seeking the preservation of ecological balance in the area. Support from men has helped the women to continue with their protest, which they saw as promoting a 'back to nature' strategy instead of an alternative and more 'progressive' type of economic development, like those taking place in the rest of the country. Yet Jain (1984) says that men did not see it as a survival issue as the women did due to 'the way the subsistence economy is organized in this area' (p. 1788).

Whether the Chipko workers intended it or not, all of the women who participated in the Chipko meetings and processions became aware of their potentialities and demanded a share in the decision-making process at the community level. Gaura Devi, an older peasant woman who faced down loggers and armed guards, has subsequently been called the Rosa Parks of the Indian environmental movement and mother of the Chipko movement. On 26 March 1974, when the large-scale felling of trees was initiated and mountain people became aware of its dangers, women under the leadership of Gaura Devi held a three-day, three-night vigil that succeeded in preventing the lumbermen from commencing their work (Jain, 1984).

Until the 1970s, *van panchayat* or community forestry management councils seemed to have been almost exclusively male due to the traditional exclusion of women from community institutions. The Chipko movement drew women out of their homes for the first time, exposing them to a new world of assertion and to the possibility of voicing and articulating their demands. The movement also had a significant impact on traditional gender relations, with men accepting women's participation in non-domestic affairs, particularly those related to the management of forest resources. Participation enabled the Mahila Mangal Dal, a women's welfare group, to effectively take control of the day-to-day management of the village forest from the male-dominated *van panchayat* council, with the help of President Gaura Devi (Sarin, 2001).

Although pleased about having gained control over the village forest, the women expressed their resentment over the fact that the men left all the forest protection work to them. They had attempted to coax the men into assisting with voluntary patrolling, but the men had refused, saying that only women needed the forest. The women also reported that when outsiders came to the village – and there were many visitors due to the contacts established during the Chipko uprising – the men pushed the women forward to talk to those outsiders (Sarin, 2001). Interestingly, the men had – advertently or not – succeeded in casting women in the role of community advocates, thus facilitating women's involvement and empowering women's voices in the movement. In 1986, Gaura Devi received the first Friend of the Environment award, known as the Priyadarshini Vrikshamitra Award. By the time of Devi's death in 1991 at the age of 66, women's concerted resistance to the felling of trees had also stopped in the Reni village in Chamoli district. The reason? The women who had set in motion the Chipko movement were now seeking alterations in their own situation.

Furthermore, unlike Mahatma Gandhi, the leaders of the Chipko movement had not at this point in time formulated their views on women's position in society. Thus, the women's spontaneous and determined participation in the movement was not constrained in any way by a pre-existing ideology. Village women had in effect been organizing their own struggles within their households and communities in order to gain greater control over their forests. Their daily lives were most intimately impacted by the quality of the forest, its proximity to the village, and the institutional structures determining forest access. The limited number of institutional structures determining forest access is key to understanding women's involvement in the Chipko movement (Sarin, 2001).

2.4 Critical Reception of the Chipko Movement

There are indeed observers who have been critical of the romanticism of some of those enthralled by the Chipko movement. Rangan (1996), for example, criticizes environmentalists inside and outside of India for being 'rapt and slavish in their adoration and assiduous pursuit of romance with Chipko's ecological reincarnation' (p. 222). She suggests that Chipko's leaders were in fact reactionary, and that their allegiance to the myth of Chipko has prevented the protest and arguments of village leaders and activists from being heard outside of the region. It has also been noted that there have also been militant calls within the local community for more development – that is, for tree-felling rather than tree-hugging.

The short-lived and ultimately inconsequential movement for the establishment of the new state of Uttaranchal, in particular, embraced a message of secular development and social justice that commanded widespread political support locally. As a result of the Chipko movement winning the support of the government and environmentalists, this wider movement has been stifled and the message of secular development and social justice has fallen on deaf ears. Furthermore, it has been argued that the success of the Chipko movement has also influenced protests against another kind of development, namely the construction of the Tehri Dam Project in Tehri Garhwal (Bandyopadhyay, 1992).

The Chipko movement of the 1970s began among people protesting against their livelihood being taken away by the Indian government from their Himalayan region. The movement was successful in changing the forest policy, but after that, members did not experience any further economic opportunities and development. Scholars such as Rangan have shown that liberation ecology – or any contemporary debate over development and the environment – must encompass the livelihood of the people as an essential part of environmental sustainability. In her book *Liberation Ecologies: Environment, Development, Social Movements*, Rangan shows in detail how the Uttaranchal movement emerged as an offshoot of the Chipko movement for a short period of time from 2000 to 2006, becoming somewhat violent when a separate state of Uttarakhand was born as a part of Uttar Pradesh. Uttarakhand included eight Himalayan districts with the goal of promoting economic development by addressing the issue of social equity in relation to the marginalized people of this region. As Rangan (1996) maintains, 'Development is a dynamic process that involves states, market and civil societies to varying degrees in actively reshaping social relations and institutions' (p. 207). She explains how the development planning in post-independence India involved a dynamic process of working and reworking among regions, sectors, and constituencies in order to achieve greater socio-economic and political equality. The Chipko and Uttaranchal movements were linked by the common issue of concerns over regional economic development. This demand was directed against regulations and policies that severely constrained local and regional economic development. Chipko's initial attempts to alter regulations and demand economic concessions were overwhelmed by its subsequent fame as a grassroots ecological movement.

Moreover, Chipko's ecological successes resulted in new environmental regulations that compounded the lack of economic opportunities and development in the region. It was this situation that gave rise to the demand for creating an Uttaranchal state, and a part of the state of Uttar Pradesh became Uttaranchal for a short period of time from 2000 to 2006. In 2006, the name was changed to Uttarakhand. The growing violence in the Uttaranchal movement expressed the frustration of some participants who realized that their protests had neither gained wider political support nor succeeded in forcing the Indian government to negotiate the issue of statehood. New social movements in the developing world, Rangan argues, are not against the idea of development but are rather an essential part of it. Thus, it is important to recognize and acknowledge that the broad notion of development – as a means of gaining access to social equality, promoting economic well-being, and fostering political recognition of marginalized communities – remains central to any liberation ecology project.

Emerging from the necessity of dealing with the social and economic problems faced by the nation, development planning was also reshaped by contingent events such as wars, droughts, energy crises, and the unexpected outcomes of its own policies. Rangan say, 'India's planners set out their policies as broad guidelines within which social and economic transformation was to occur, reforming their strategies every five years in response to problems that were entirely new, unforeseen, or stubbornly persistent, despite reassessment and reformulation' (p. 211). Development planning in India was a multifaceted process involving states, markets and civil societies, each of which had been active in varying degrees in reshaping social relations among different classes of people and social institutions. To simplify, the main point of Rangan's criticism is that those scholars who have regarded the Chipko movement as not being concerned with the economic model of growth but as being based on the Gandhian idealistic model have ignored the complexity of the context in which the movement emerged and advanced.

Mawdsley is similarly critical of Vandana Shiva and Ramachandra Guha's analysis of the Chipko movement as an ecofeminist and peasant movement respectively. In her 1988 article, 'After Chipko: from Environment to Region in Uttaranchal', Mawdsley argues that neo-populist theorizing about Chipko has weaknesses. One of the weaknesses is that in analyzing the Chipko protests, the political and economic contexts have been ignored by highlighting the ecofeminist and eco-centric portrayals of the movement. The fact that very little is left of the Chipko movement in the place where it first originated is not analyzed. The DGSS (Dasholi Gram Swarajya Sangh), a small industrial corporation with Gandhian overtones based in Gopeshwar under the leadership of Chandi Prasad Bhatt, ran a turpentine unit, manufactured agricultural implements, and organized demonstrations against the sale of liquor, untouchability, and the forest contractor system (Guha, 1989; Mawdsley, 1998). In 1973, DGSS came into conflict with the Forest Department when it allowed a large sports equipment company, Symonds, to cut hundreds of trees when DGSS was not allowed to cut a few trees from the same forest. DGSS decided to stop Symond's contractors by intervening between the axemen and the trees, thus giving rise to the movement's famous name Chipko, which means 'cling to' in Hindi. After successfully preventing tree-felling in Mandal forest, Chandi Prasad Bhatt and others decided to alert the villagers around Phata-Rampur, about 80 km (49.7 miles) away, that the forest around them would be cut by the same company and offered help to defend them. Mawdsley's point is that it was by moving beyond their immediate local needs to embrace a wider spatial and temporal perspective that Chipko was born as a meaningful social movement with regional implications. Related to this is another point: that the grassroots and spatial diversity of the movement has not been recognized by neo-populist theorizing. The struggle in the Mandal village was all about access to raw materials for small-scale industrial use. By contrast, in the Reni village where the celebrated incident of 1974 occurred, women were prominent in protecting their forests from the contractors, and the industrial opportunities here were much more limited. Reni is a remote village close to the Indo-Tibetan border and further away from a road of poor quality. It is here that a large number of women were responsible for the work related to the forest. Most analysis of the Chipko movement fails to take into consideration this diversity. Both the marginalization and the degradation of an ecological base on which the local people depended played equally important roles, but the more popular accounts of the Chipko movement downplay the economic demands of the movement and highlight only the environmental perspective.

Mawdsley points out that the short-lived Uttaranchal movement – which became a mass movement in July 1994 – confronted the state and demanded a separate Uttaranchal state. As Mawdsley (1998) explains, 'The main grievance being articulated is that the economic and developmental marginalization of the hill area is due to the fact that the *plains*-based planners in the distant State capital of Lucknow are unable (as well as unwilling) to understand the development needs of the *hill* population, environment and economy' (p. 5). Only 4 percent of the state of Uttar Pradesh's population lives in the hill area, and their interests are completely ignored. This is why the Uttaranchal activists believed that with the formation of a separate Uttaranchal state, the hill people's political voice would be heard. Mawdsley's point is that the neo-populist account of the Chipko protest does not take into consideration the complexities of the lack of control over local resources, the competition over national versus local needs, the environmental concerns, and the local critiques of development planning and administration. Neo-populist theories put forth a different kind of understanding of gender relations, development, and the environment.

Mawdsley is critical of Shiva's analysis of the Chipko protest which depicts women as the only representatives trying to re-establish an ecological and traditional harmonious relationship with nature, and in the process overlooking the fact that many men were committed to the different strands of the Chipko movement. Also, from the field study and the interviews that Mawdsley had conducted with the hill people, she discovered that both men and women in the hill were interested in finding preferably government jobs with a secure income. Mawdsley does not deny the agro-environmental concerns of many women, but it is important to note that they were equally concerned with 'other ways of winning livelihood in Uttaranchal, including the commercial exploitation of the forests – something that Shiva tends not to acknowledge' (1998, p. 10).

Mawdsley is also critical of Guha's portrayal of the resistance as one of peasant versus state. She argues that this might have been true in the 1970s, but in the 1990s the Uttaranchal movement's 'regional struggle was the *capture,* not the *rejection,* of the state and thus state power' (1998, p. 11). Mawdsley shows that the neo-populist interpretation gives a one-dimensional and simplistic account of the Chipko and Uttaranchal movements. Thus, by not taking into account the complexities of the movement, neo-populist theorizing has done injustice to the movement.

2.4.1 Questions about Chipko's Popularity and Success

Although the Chipko movement's achievements in terms of some of its objectives were modest, given its wide following and success, it can be regarded as a watershed phenomenon in the history of Indian environmentalist activism. The Chipko protest, in fact, kindled the environmental aspects of development and gave rise to numerous conflicts and protests over natural resources and ecological issues. However, none of the many forest-based movements during the last two decades in India have attracted as much public support or influenced state policies as Chipko did. This may be attributed to three important aspects of the Chipko movement (Reddy, 1997).

First among them is the close link between the livelihood of the local people and the nature of the movement. The local people considered Chipko a fight for basic subsistence denied to them by the institutions and policies of the state (Reddy, 1997; Guha, 1989). In addition, the contribution of women to their households' subsistence and their participation in the anti-alcohol campaign led to the overwhelming support of women, which was unique to the Chipko movement (Guha, 1989). Kumar (1993) notes that the Uttarakhand Sangharsh Vahini formed in 1977 was also active in the Chipko movement by going from village to village to spread and collect information. The problem of alcoholism among men was confirmed by this group. She says that the anti-alcohol movement was closely linked to the Chipko movement in Garhwal and Kumaon because the men's alcohol problem meant that the women suffered, as they didn't have enough money to feed their family. 'According to Uma Bhatt, a sustained agitation against the sale of alcohol had taken place from 1965-1971 in Uttarakhand' (Kumar, 1993, pp. 186-187). As aptly described by Guha (1989), the 'private' image of Chipko is more of a peasant movement, while its public profile is that of an environmental movement, which gave rise to the myth of Chipko. Furthermore, the public face of the movement as a women's movement is strongly pronounced (Omvedt, 1993). It is interesting to note that it was in the later stages when Chipko ceased to go beyond environmental concerns – that is, limiting itself to the protection and conservation of trees – that problems started surfacing.

The second aspect accounting for the popularity of the Chipko movement has to do with the nature of the agitation. Unlike other environmental movements, the Chipko adhered very strictly to the Gandhian tradition of freedom struggle: nonviolence. In Guha's own words,

> there is the veneer of Gandhianism with which Chipko is cloaked, a matter of some embarrassment for a state claiming to be the rightful successor of the freedom struggle and upholding Gandhi as the Father of the Nation. In this manner Chipko has, knowingly or unknowingly, successfully exploited the ambiguities in the dominant ideology of the Indian state. (1989, p. 177)

Third, the simplicity and sincerity of leaders such as Sunderlal Bahuguna and their access to national leaders of the caliber of Prime Minister Indira Gandhi and other politicians and officials also helped to boost the success of the movement.

Despite its popularity and success, the Chipko movement is still considered to be incomplete and modest because it limited itself to ecological issues of protecting trees while at the same time neglecting local people's requirements (Reddy, 1997). It should be mentioned that the Chipko movement had articulated six distinct demands and that only one of those demands concerned the complete termination of the commercial cutting of trees. The other demands were to: (i) reorganize traditional rights on the basis of meeting the minimum needs of the local population; (ii) turn arid forests green with the help of active village participation and increased tree cultivation; (iii) form village committees to manage forests; (iv) develop forest-related home-based industries and supply those industries with raw materials, money, and technology; and (v) give priority to local varieties in afforestation based on local conditions and requirements (Mukul, 1993). Recent evidence shows that due to the neglect of eco-development linkages (markets, techniques, forest-based industries), the objective of ameliorating the conditions of the local people in these regions remains a distant dream. If the situation continues – that is, if threats to the livelihood of the local people are not prevented or at least mitigated – this may jeopardize the achievements of the Chipko movement and contribute to the movement's loss of support. Indeed, the shutting down of local industries has caused most entrepreneurs – who were initially attached to the movement – to feel alienated from it (Mukul, 1993).

Rangan points out that the protest against deforestation in the Himalayan region was heterogeneous. In the eastern districts of the Garhwal Himalayas, the protest took the form of hugging (Chipko) the trees to prevent the migrant workers hired by the forest contractors from cutting the trees. The main aim of the members of the Chipko movement was to get access to small-scale forest use and to pressure the local government to provide them with developmental assistance in their region. There was another group that demanded that the large-scale felling of trees by forest contractors be stopped. Yet another group wanted to promote a forest-labour cooperative so that their right to access the forest could expand and in turn help the local communities. Others opted to demand a total ban on the export of raw materials from the region. It is important to note that, on the one hand, there were groups demanding higher wages for forest labourers, and on the other hand, there were those inspired by Gandhian ideals who were interested in promoting timber subsidies and the supply of other forest commodities at a discount rate to help the local population. It was clear that the negotiation with the Forest Department had failed. Forest officers claimed that they had to follow the law and comply with the production target set by national and state level plans. The local community saw the small-scale felling of trees as a threat, since local contractors were not given the opportunity to participate in the commercial felling of trees and since this policy would lead to a monopoly and thus inefficiencies (Rangan, 1996).

Faced with this gridlock in negotiation, Sunderlal Bahuguna appealed directly to the Indian government instead of negotiating through the state of Uttar Pradesh. Bahuguna, a local forest contractor, and a spokesperson of the Chipko movement put pressure on the central government to interfere, arguing that it threatened the national security of the Himalayas. Bahuguna emphasized the importance of the forest in terms of national security and the defense of the border. Chipko members expressed their outrage against commercial deforestation driven primarily by the profit motive. They explained how commercial extraction took away the simple needs of the local peasants. As Rangan (1996) explains,

> Bahuguna's narrative cast Chipko's followers as victims of natural disasters, the state, and the market. Floods, poverty, emigration, and women's suffering and daily struggle to collect fuel and fodder for their households were described as inevitable consequences of timber extraction by forest contractors. (p. 216)

Bahuguna's plea led to the immediate promise from the central government of a ban on tree-cutting for commercial purposes in the Himalayan region. This success gave rise to media coverage of the movement. In turn, scholars and activists not only in India but also outside of India supported Bahuguna's cause, and he came to be seen as the natural leader of the Chipko movement, the voice of a grassroots struggle that sought to protect the simple, peasant ways of life and to restore harmony between humans and nature in the Himalayas (Rangan, 1996).

2.5 Conclusion

As one of the first grassroots movement in India, Chipko emerged in the 1970s as a reaction to the destruction of the forests of the Himalayan Garhwal. This chapter has relied on a review of expert literature on the subject to demonstrate how the movement actually took place as a social movement of the peasants in the state of Uttarakhand. When it spread to other regions of India and eventually the world, scholars began to view the Chipko movement as an environmental movement embraced by women. Though some scholars have argued that the movement has not been appraised in all its complexities because a few aspects have been highlighted while others have been ignored, the Chipko movement's contribution to the discourse on the environment, on development, and on sustainability by including the voice of marginalized people in India as well as around the world has nonetheless been remarkable.

This chapter has explored the distinguishing features of this movement in connection with the leadership of Chandi Prasad Bhatt and Sunderlal Bahuguna and in relation to the role played by women. Some critics have called this perception and depiction of the Chipko movement an instance of neo-populist theorizing, and others have called it the 'myth of Chipko'. As was noted, both Chandi Prasad Bhatt and Sunderlal Bahuguna have had extensive training in the Gandhian discipline of holding onto truth at all cost, which is the Gandhian method of *satyagraha*. This principle of nonviolence has been the foundation of the Chipko movement. Their followers, mostly women whose livelihood was taken away, also used the method of hugging trees to stop the contractors from felling the trees and succeeded in forcing the Indian government – under Prime Minister Indira Gandhi – to impose a fifteen-year ban on tree-felling in the state of Uttarakhand in 1980. The question of how and why Chipko accomplished its goal has been analyzed critically. It has been argued that the Chipko movement was mainly concerned with issues of social justice arising from the disproportionate use of forest resources by the people and groups in positions of authority and power.

I turn next to another comparable Indian environmental movement that emerged in connection with the issue of dam-building: the Narmada Bachao Andolan (Save the Narmada Movement). As we shall see, this movement was also based on the Gandhian method of *satyagraha* and was embraced by indigenous people who had been unjustly displaced.

Works Cited

Bahuguna, Sunderlal (1990). 'The Chipko Movement'. In Lina Sen (ed.), *A Space Within Struggle: Women's Participation in Peoples Movements.* New Delhi: Kali for Women.

Bandyopadhyay, Jayanta (1992). 'Sustainability and Survival in the Mountain Context'. *Ambio*, 21, 4, 297-302.

Gosling, D.I. (2001). *Religion & Ecology in India & Southeast Asia.* London: Routledge.

Guha, Ramachandra (1989). *The Unquiet Woods: Ecological Change and Peasant Resistance in the Himalaya.* Delhi: Oxford University Press.

—, and Juan Martinez-Alier (1997). *Varieties of Environmentalism: Essays North and South.* London: Earthscan.

Jain, Shobhita (1984). 'Women and People's Ecological Movement: A Case Study of Women's Role in the Chipko Movement in Uttar Pradesh'. *Economic and Political Weekly*, 19, 41, 1788-1794.

James, George A. (2000). 'Ethical and Religious Dimensions of the Chipko Resistance'. In Christopher Key Chapple and Mary Evelyn Tucker (eds.), *Hinduism and Ecology: The Intersection of Earth, Sky, and Water.* Cambridge, MA: Harvard University Press. 470-499.

Kumar, Radha (1993). *The History of Doing: An Illustrated Account of Movements for Women's Rights and Feminism in India, 1800-1990.* London & New York: Verso.

Mawdsley, Emma (1998). 'After Chipko: From Environment to Region in Uttaranchal'. London: Birkbeck ePrints. https://eprints.bbk.ac.uk/218/1/mawdsley.pdf (accessed 8 June 2018).

Misra, Shalini (2007). 'Spirituality, Culture and the Politics of Environmentalism in India'. *The Journal of Entrepreneurship*, 16, 2, 131-145.

Mukul (1993). 'Villages of Chipko Movement'. *Economic & Political Weekly*, 28, 617-621.

Omvedt, Gail (1993). *Reinventing Revolution: New Social Movements and the Socialist Tradition in India.* London: East Gate Book.

Rangan, Haripriya (1996). 'From Chipko to Uttaranchal'. In Richard Peet and Michael Watts (eds.), *Liberation Ecologies: Environment, Development, Social Movements.* London and New York: Routledge.

Reddy, Ratna (1997). 'Discussion Paper 64: Environmental Movements in India: Some Reflections'. www.sai.uni-heidelberg.de/intwep/fia/DISKUS64.htm (accessed 16 June 2018).

Sarin, Madhu (2001). 'Empowerment and Disempowerment of Forest Women in Uttarakhand, India'. *Gender, Technology and Development*, 5, 3, 341-364.

Shiva, Vandana, and Jayanta Bandyopadhyay (1986). *Chipko: India's Civilizational Response to the Forest Crisis.* New Delhi: INTACH.

3 Narmada Bachao Andolan (NBA): Save the Narmada

Abstract
After tracing the history of dam-building, specifically the Sardar Sarovar Project in the Narmada river which traverses the three states of Gujarat, Maharashtra, and Madhya Pradesh, this chapter analyzes the Narmada Bachao Andolan protest movement which started in the 1980s and was led by Medha Patkar and others following the Gandhian method of *satyagraha*. The NBA took the SSP to the Indian Supreme Court with the help of the World Bank, leading to the suspension of the project for a short time. The NBA is continuing its efforts to obtain a proper rehabilitation policy for the indigenous population (*adivasi*) displaced by the SSP, as the government has infringed on the right to survival of the people living around the dams. The NBA has made a global impact by launching a dialogue about biodiversity and sustainability.

Keywords: Sardar Sarovar Project (SSP), Medha Patkar, sustainable development, gender justice, *adivasi*

The *Narmada Bachao Andolan* (NBA) or 'Save the Narmada' movement originated in the 1980s as a protest against the building of dams in the Narmada River, a major waterway flowing across the Indian states of Madhya Pradesh, Maharashtra, and Gujarat. The movement was led by a woman, Medha Patkar, and was embraced by poor indigenous people who were displaced due to the dam-building. Whereas the Chipko movement a decade earlier had focused attention on preserving access to forest resources by local people, the NBA insisted that those people whose rights to local resources cannot be protected due to development – of a dam, for example – should be compensated as well as resettled. In this way, economic justice is served. Dam-building, like forest clear-cutting, has a long colonial and postcolonial

Mallick, Krishna, *Environmental Movements of India: Chipko, Narmada Bachao Andolan, Navdanya*. Taylor & Francis Group, 2021.
DOI: 10.5117/9789462984431_CH03

history in India. By 1999, India had become the world's third largest dam-builder, trailing only China and the United States.

The specific target of the NBA's protests was the development of the Sardar Sarovar Project (SSP), the second largest of 30 very massive dams that were part of a larger plan for 3,000 dams in India. Today, the SSP has one of the largest canal networks in the world. The primary rationale for building the dam was to provide irrigation and drinking water, but as a multipurpose dam and canal system, it also produces electrical power. The submerging of a large area of forests left scores of tribal people deprived of food, fuel, and medicine. NBA protesters railed against the displacement of the area's *adivasis*, or indigenous people, and the injustice of failing to adequately resettle and compensate the community. In doing so, the NBA unified social movements across India in opposition to social injustice, political exclusion, and environmental degradation. As Misra (2007) points out in her article 'Spirituality, Culture and the Politics of Environmentalism in India', the NBA was successful in claiming that although the tribal people might not have a legal right to the land, they had a right to be compensated for losing their land. What made the situation extra controversial was that the dam was situated in the state of Gujarat, which was to benefit the most from its operation, but the dam's submergence primarily affected the state of Madhya Pradesh and, to a much lesser extent, the state of Maharashtra.

The NBA movement is unique in that it has emphasized both the 'red' and 'green' ideologies of nonviolent resistance, according to Basu and Silliman (2000). The 'red' ideology refers to the movement's battle for social and environmental equity for the most marginalized groups of Indian society, the non-Hindu tribal population. It does not focus on the upper-class urban Hindus who are the primary focus of the central government of India and the state governments of Gujarat and Madhya Pradesh. The ecological destruction caused by the building of dams refers to the 'green' ideology of the movement (Misra, 2007). Alongside the intellectual leadership of the Gandhian environmental movement, the NBA developed a strong critique of the role of the state and the 'modern Western' scientific project of dam-building, arguing that science in the service of large-scale capitalist and bureaucratic projects aimed at the appropriation of the natural commons is violent and against people, livelihoods, and cultures. The Gandhians advocate replicable, small-scale, decentralized, democratic, and ecologically sustainable activities. Of course, the Gandhian environmental movement does not profess to be an anti-science movement; rather, its advocates oppose the practice of science that dismisses ultimate values and cultural goals.

They reject the idea that the plurality of lifestyles and knowledge systems must give way to the merciless machinery of progress (Rajan, 2005).

3.1 The Common Good in a Cost-Benefit Analysis

Central to the Indian government's claim that the mega-dams were essential to economic development was the assertion that the SSP would benefit millions of people living in the Narmada River valley. The gain, they said, would outweigh any potential human or environmental costs. This narrative of the 'common good' has been emblematic of the government's stance throughout the many controversies generated by the project. The 'common good' approach reflects the dominance of a 'balancing' or 'cost-benefit' approach over one that puts human rights at the centre of the debate (Kingsbury & Lustig, 2006).

Today, there are more than 45,000 large dams in 140 countries (Agoramoorthy and Hsu, 2008) that generate one-fifth of the global power supply. The stored water from these dams allows farmers to produce about one-sixth of the earth's total food production (Leslie, 2005). Since the completion of the Hoover Dam in the Mojave Desert in 1925 – the world's largest dam at the time and a technological marvel – policymakers around the world have held out hope that building dams will foster the growth of glamorous cities such as Las Vegas and Los Angeles (Agoramoorthy and Hsu, 2008).

Not all dams are economically advantageous. Most dam development projects are viewed in terms of their costs and benefits. This is true for micro-level case studies of specific dam projects as well as for more global evaluations. By this logic, the benefits that dams create through controlled floodwaters, increased irrigation, and electricity should be balanced against the negative ecological and social impacts on both downstream and upstream populations and environments. About 57 percent of dam projects funded by the World Bank – a major financier of large dams – have been judged as unprofitable in economic terms, with internal rates of return of less than 10 percent. More than three-quarters of the World Bank's dam projects returned less than expected value at appraisal (Whitehead, 2003).

What a cost-benefit analysis does not take into consideration is the fact that those who benefit from dams and those who pay the costs are always two very different groups of people in terms of their economic, legal, cultural, and political vulnerabilities. To compare the costs and benefits side by side is to compare apples and oranges, because they are two incomparable entities. In the process, the lived social relations of people whose lands are being submerged do not balance out with the relations of production and exchange with which people become intricately attached. Yet even those studies that stress the high social and ecological costs of dams are not willing to give up the cost-benefit analysis, and so the debate continues (Whitehead, 2003).

Instead of large dams, the NBA proposed energy and water conservation based on improved dry farming technology, watershed development, small dams, lift schemes for irrigation and drinking water, and improving the efficiency and utilization of existing dams (Jensen, 2004; Duara, 2015). Ultimately, with the NBA making headlines, hundreds of nongovernmental and intergovernmental organizations worldwide were prompted to come into action, calling for 'an independent comprehensive review of all [World] Bank-funded large dam projects to establish the actual costs, including direct and indirect economic, environmental, and social costs, and the actually realized benefits of each project'. (as cited in Ramachandra 2006, p. 277; Manibeli Declaration (accessed 28 May, 2020) http://www.rivernet.org/manibeli.htm)

The independent review was conducted by the twelve commissioners of the World Commission on Dams (WCD), which included NBA founder Medha Patkar. The commission's final report published in November 2000 found that large dams are generally unfavourable as a means for development due to the high levels of associated environmental and social risk. The WCD report also noted that most large dams fall short of their intended targets for irrigation, power generation, or water supply (World Commission on Dams, 2000). In terms of the impact of dams on people, the report noted that the 'negative effects were frequently neither adequately assessed nor accounted for', although the 'range of these impacts is substantial, including on the lives, livelihood, and health of the affected communities' (WCD, 2000). Moreover, enumeration and compensation programmes for the displaced are 'often inadequate' (Ramachandra, 2006).

Finally, the WCD report stressed the need to fully and accurately consider the costs to society and the environment when assessing current and future dam projects, particularly through consultation with affected parties (Ramachandra, 2006). This fueled the NBA movement and ultimately led to greater oversight of dam-building worldwide. As Ramachandra points out in her 2006 article 'Sardar Sarovar: An Experience Retained', that local struggle and international opposition against large dam projects – coordinated in the beginning by the International Rivers Network and the NBA – had galvanized a movement that ultimately resulted in the formation of World Commission on Dams (WCD) due to the success of the Independent Review.

The twelve commissioners of WCD, including Medha Patkar, founder of NBA, released a report revoking the funding for the SSP.

3.1.1 A Recursive History of Dam-Building

Dam-building, a means of controlling nature, is simultaneously a grab for power, writes Basu in his article 'Scale, Place and Social Movements: Strategies of Resistance along India's Narmada River'. Basu notes three flashpoints in this global trend, referring first to studies on dam and canal building during British colonial rule that describe how social inequalities were created through water control projects (Gilmartin, 1995; Whitcombe, 1972). Second, Basu confirms that postcolonial India expanded the damming of rivers as part of a national policy of industrialization (Rangachari, 2006; Dharmadhikary, 2005; Basu, 2010). The third key moment in the history of dams is the more recent opposition to them, which Basu writes, has been a central platform of the U.S. environmental movement since its birth in the 1970s.

This advocacy has given strength to other anti-dam movements (Worster, 1985). The WCD (2000) reports the extent to which anti-dam struggles are dominant within international development agencies (Dubash et al., 2001). As Basu (2010) notes,

> The movement against the Narmada dams can be situated within all three of these historical moments, since the project embodies colonialist visions of social control, as well as public ideas of welfare and national interest. The NBA has become a visible symbol of an international anti-dam movement. (pp. 101-102)

Coincidentally, Worster's analysis (1985) of water control in the Western United States also draws a connection between control over nature and control over society. He writes that the British dam-building model was used by American engineers because it was easy to follow. Models of dam-building thus originated in colonial contexts and travelled globally as part of a wide-ranging effort to exert power through control over nature (Basu, 2010).

Proponents of dams have not only pointed to the agricultural and industrial developments made possible by irrigation and hydroelectricity but also emphasized the moral imperative that dam-building would lead to opportunities for public employment, as it did during the 1930s Great Depression in the United States (Brooks, 2006). The Tennessee Valley Authority (TVA) has been mentioned in this regard as a model for contemporary dam-building in India. The imitative trend of dam-building has come full circle, from British India's colonial models being borrowed by the United States and U.S. models subsequently being implemented by India (Basu, 2010).

To this day, anti-dam protestors strive to achieve the goal of sustainable development. There are many definitions of sustainable development, including the following landmark definition that first appeared in the well-known Brundtland report of 1987 (World Commission on Environment and Development), 'Development that meets the needs of the present without compromising the ability of future generations to meet their own needs' (as cited in De 2014, 2). According to De (2014), 'people concerned about sustainable development have long suggested that meeting the needs of the future depends on how well we balance social, economic, and environmental objectives or needs when making decisions today' (p. 2). By analyzing how the Sardar Sarovar Project was started and how opposition to the project took shape with the formation of the NBA, it becomes clear that the failure to implement ideal sustainable development gave rise to many controversies.

3.2 Regional Tensions from the Start

Prime Minister Nehru inaugurated the SSP in 1961, but construction did not begin immediately, as the three states involved – Gujarat, Madhya Pradesh, and Maharashtra – could not initially agree on the allocation of the Narmada River's waters. Gujarat filed an official complaint to the central government in 1968 under the Inter-State Water Disputes Act. More than a decade later, in December 1979, the Narmada Water Disputes Tribunal (NWDT) published the NWDT 'Award' in the newspaper *Gazette of India*, explaining the basis on which work had been carried out on the dam until then. The tribunal decided that the length of the full reservoir level would be 455 feet (138.68 metres) and determined the shares of Narmada water each state was entitled to use, allocating among them the shares of power to be produced by the dam.

The 1979 NWDT Award dealt with the problem of displacement and resettlement. It acknowledged that all lands under private ownership below 455 feet should be acquired by the states, and that all people living on these lands should be rehabilitated. The Award was progressive insofar as it provided that land should in principle be given in exchange for land compulsorily acquired (Patel, 1997). While the number of families affected by the project was estimated to be 6,147 in 1979, the official estimate of the three state governments of Gujarat, Maharashtra, and Madhya Pradesh is now 41,000 families (205,000 people in total). Neither estimate counts the approximate 157,000 people displaced by the canals nor any of the people displaced because of compensatory afforestation or as a result of the project construction colony (Cullet, 2001, pp. 973-974).

The fact that the river flows between the states provided the legal basis for the NWDT to set up an inter-state administrative authority known as the Narmada Control Authority (NCA), which was to facilitate the implementation of the Award. The NCA employed the government of India, state engineers, and civil servants to coordinate and monitor the resettlement of people whose homes would be flooded by the reservoirs as well as the financing and construction of resettlement villages. The central Indian government retained some control over the details of the projects through the Central Water and Power Commission, the Planning Commission, and the Ministry of Environment and Forests, each of which had authority to grant or withhold clearance for proceeding with various aspects of project construction (Peterson, 2010; Cullet, 2001). Finally, the World Bank sanctioned a loan of $450 million in 1985, when the Sardar Sarovar Dam gained international prominence (Narula, 2008).

The three state governments took responsibility for their own stretches of the river in the detailed design, construction, operation, and maintenance of the Narmada River system (Peterson, 2010). Gujarat and Madhya Pradesh established public corporations for this endeavour: Sardar Sarovar Narmada Nigam Limited (SSNL) in Gujarat, and Narmada Valley Development Authority (NVDA) in Madhya Pradesh. The NWDT Award did not end the regional arguments entirely: Gujarat and Madhya Pradesh continued to disagree about the height of the dam to be built at Navagam, now named Sardar Sarovar Dam, and hence the size of the reservoir to be created. However, those arguments were soon overshadowed by controversies over treatment of the families that were ousted and the environmental impacts of the entire Narmada River project (Peterson, 2010; Cullet, 2001).

3.2.1 Amid Unrest, NGOs Align to Form NBA

While the governmental groups were organizing, resentment was brewing among the communities in the Narmada River basin. The social stratum of those who were displaced is not unique to India or the rest of the world. Of the 200,000 to 250,000 people who have been or are in the process of being displaced in India, 60 to 70 percent belong to 'scheduled tribes' that live primarily by pastoralism – that is, subsistence-oriented, slash-and-burn agriculture and/or hunting and gathering. These indigenous groups are officially recognized in the Indian Constitution. They are economically and socially speaking borderline people: marginalized by their numbers, their distinctive modes of production, and their relative geographical and social isolation from mainstream society (Whitehead, 2003).

Displacement brought on by large infrastructure projects remains an unresolved issue in the quest for development. Post-colonial development policy and planning have largely followed a utilitarian view that allows for millions to be displaced in the interest of the good of the majority. The forced uprooting of whole communities was considered the 'cost' of development that had to be borne in the name of the overarching national interest. By linking the notion of 'eminent domain' to the colonial Land Acquisition Act of 1894, the Indian government has acquired the power to appropriate private property under the justification of 'national purpose' (Mehta, 2008).

Another sharp counterpoint to arguments in support of the SSP is presented by Penz, Drydyk, and Bose in their book *Displacement by Development*. The chief questions that motivate these authors are *who* benefits from such dam-building projects and at *what* cost? The approach promoted in this work is labeled 'development ethics', an interdisciplinary viewpoint that seeks to evaluate issues such as displacement from a very different perspective.

Responding to the unrest, local opponents, environmental activists, and professionals from the academic, scientific, and cultural worlds founded a cluster of non-governmental organizations (NGOs) that gained strength in the late 1980s when they joined up to form the Narmada Bachao Andolan. Conflict between the NBA and the government of Gujarat started with the improper rehabilitation of people dislocated by the Sardar Sarovar Dam. Arundhati Roy (1998) describes the actions taken by the people affected:

> In September 1989, more than 50,000 people gathered in the valley from all over India to pledge to fight 'destructive development'... One year later, on 28 September 1990, thousands of villagers made their way on foot and by boat to a little town called Badwani, in Madhya Pradesh, to reiterate their pledge to drown rather than agree to move from their homes. (pp. 37-38)

Alongside demonstrations of *satyagraha* that included the potential for drowning (as Roy notes) or fasting by Medha Patkar and others, the NBA filed a petition in the Supreme Court of India on behalf of those tribal people who had been ousted and others who were directly and indirectly affected by the SSP. This writ petition, filed in 1994, challenged the construction of the project on the grounds that it was 'unviable from a social, environmental, technical, economic and financial point of view… [and] was not in the national interest' (Supreme Court of India, Writ Petition [Civil] 319 of 1994: NBA v. Union of India). Following the writ, the Supreme Court of India ordered a halt in construction in 1995 but subsequently allowed construction to resume in 1999. The Court issued an opinion on the future of the dam on 18 October 2000, immediately approving an increase in the dam height, conditioned upon approval of the Relief and Rehabilitation and Environmental Sub-groups of the Narmada Control Authority (Ramachandra, 2006).[1]

The desire for economic development was understandably powerful, but doubts and suspicions lingered despite the majority ruling issued in 2000 by Justice Kirpal and Justice Anand of the Supreme Court of India, which had carefully assessed a revised plan and pronounced itself in favour of the project (Blewitt, 2015; see also Supreme Court of India Judgement 2000).

3.2.2 National and Global Ramifications

The NBA activists' fight has been for indigenous people's right to be compensated for their displacement as a result of dam-building and subsequent natural disasters, like flooding. With their nonviolent actions, they sparked a debate about World Bank funding both in India and around the world, which led to the dam-building plans for the SSP being revised. Nationally, the NBA not only opposed the dam but proposed various development alternatives, including decentralized methods of water harvesting. Internationally, the movement led the charge in demanding World Bank accountability for its involvement in a project that threatened to harm millions. Their campaign led to the creation of a World Bank commission in 1991 to independently review the project. The independent review, chaired by former high-ranking United Nations official Bradford Morse, ultimately recommended that the Bank withdraw its funding from the project (Narula, 2008). Narula notes that the citing of human rights concerns reached far beyond the Sardar Sarovar Dam. The Morse report was blunt in its assessment of the situation. Focusing on the participation of those most directly affected, the review concluded that 'unless a project can be carried out in accordance with existing norms of human rights – norms espoused and endorsed by the Bank and many borrower countries – the project ought not to proceed' (cited in Narula, p. 353; Morse & Berger, 1992, p. 358). The World Bank ceased to support the project the following year.

1 See the order of the Supreme Court in the Narmada Case (accessed 9 February 2018) http://www.narmada.org/sardar-sarovar/sc.ruling/nba.comments.html).

Furthermore, Narula points out that the controversy surrounding the dam led directly to the creation of the World Bank Inspection Panel in 1993. This was a milestone for the human rights movement and the first mechanism established to enable local groups to challenge World Bank projects. Before the Panel's creation, local groups had no formal way of challenging development schemes conceived of and financed by national and international capitals where their voices were seldom heard. Despite its mixed record, the establishment of the Panel represented 'a major milestone in integrating international human rights norms into the practice of development aid' (Narula, 2008, p. 353).

Although the World Bank's withdrawal in 1993 may have been a victory, it came at a huge cost to the Narmada campaign. Seven years later, the World Commission on Dams created by the World Bank and the World Conservation Union issued a report about the aftermath of the World Bank's withdrawal of support and funding (WCD, 2000). It concluded that although the World Bank's action was a triumphant symbol of the power of mass mobilization, it had the unwanted effect of reducing the Indian government's accountability to the outside world. It also removed a body that had the obligation and ability to hold the project to a higher set of standards than the Indian government would have adhered to on its own. Narula notes that World Bank's 'withdrawal was a mixed blessing', (2008, p. 373) as the Indian government progressed with the Narmada Project in full force. She concludes,

> the NBA has contributed to an awareness of the need to approach environmental and human rights issues in tandem – a lesson long known to Indian activists and one that is increasingly apparent to actors in the international scene who urgently battle environmental degradation and its acute impact on marginalized populations. (p. 383)

3.3 Gender and the Narmada Case

The social impact of large dams has been documented, especially with respect to involuntary resettlement and displacement (Scudder, 1995; Cernea & Guggenheim, 1993). Very little is known, however, about how gender and dams are correlated. When impact assessments of large dams are conducted, gender has been one of the missing links. Women and men are not homogenous beings but instead are differentiated according to their varied social locations. As large dams affect men and women in different ways, it is necessary to analyze the gendered dimensions of the costs and benefits of large dams. Feminist literature has pointed out the different ways in which the genderless categorization of the community, the state, and its institutions have concealed the complexities unfolding in everyday exchanges between women and men (Elson, 1998; Boserup, 1970; Kabeer, 1994; Adams, 2000).

The social, economic, and environmental consequences of large dams are more widespread than those associated with other infrastructure projects because large dams impose substantial changes over time and space in the ecosystem as well as in social, economic, and cultural structures. The far-reaching effects can be felt in the environment, social organization (including family, community, and kinship networks), natural and financial resources, infrastructure development, and consumption and production processes. As Mehta and Srinivasan (2000) argue, the changes in social organization, in particular within the family and the community, affect genders differently.

The marginalization of gender concerns results in a set of misleading conclusions that are often at variance with both the realities of women and men and the specific ways in which large dams affect both genders (Mehta & Srinivasan, 2000). In her book *Displaced by Development* (2009), Mehta argues that displaced women are often caught in a double bind. Male and gender biases negatively affect displaced women in two ways. First, the widespread nature of male bias in Indian society leads to gender inequality in terms of unequal resource allocation and distribution, and it legitimizes the silencing of women's interests. Second, biases within state institutions, structures, and policies help perpetuate these societal inequalities. Mehta notes that 'At the psycho-social level, since women's lives are largely centred around the domestic realm and the family, they must have fewer outlets than men to cope with vulnerability, insecurity, and so on due to their restricted mobility' (2009, p. 25).

Communities that were displaced by the building of the dam cannot be reconstructed identically. It is well known in several resettlement villages that a few village headmen and their wives benefited from the relocation because they received prime irrigated land and achieved a new social status in the process. The same individuals are now acting as power brokers between the community and the officials (Mehta, 2009). By contrast, the new situation was less than ideal for young married *adivasi* women who, unlike many caste Hindu women, are encouraged to maintain strong links with their native homes. For generations, this had been facilitated by their marital homes being within walking distance of their native homes. With their old home and networks submerged as a result of the dam, their well-being – which was inextricably linked to the community they had grown up in – was impaired. And this facet of well-being cannot be rebuilt easily. Similarly, resettlement has exposed women to new forms of violence and insecurities that, due to blatant gender discrimination and patriarchal biases, never get addressed (Mehta, 2009). Mehta points out the need to focus more explicitly on the rights of displaced people, compared to their risks involved, as 'some risks are difficult to capture due to their intangible and non-material nature...' (p. 8).

When people are faced with displacement, they resist because their rights to livelihood, dignity, shelter, development, participation, and many other rights are denied. Mehta argues that 'Gender justice can be achieved if there is a conscious effort to move towards emancipatory justice' (2009, p. 29). This kind of justice can be achieved only when society moves beyond just the welfare of displaced women. Emancipatory justice is bound to face resistance from the local communities and state institutions, Mehta claims, due to the aforementioned 'double bind' that displaced women face.

With urban female activists like Medha Patkar acting as role models for these women, gender justice is about the actual gains that are accumulated. The displaced population of the Narmada project have sown the seeds for change by protesting. If simultaneous change takes place with regard to the policy process around R & R, there is hope that gender justice will gradually be realized both in policy and in practice (Mehta, 2009).

Despite the controversy surrounding the SSP having brought race, ethnicity, and empowerment issues to the forefront, the centrality of gender – in the construction of race and ethnicity and, more specifically, in resource management patterns and policies – has largely been ignored, as Kurian (2000) shows. Having conducted interviews and field work among both pro-dam and anti-dam groups between 1993 and 1996, she concludes that the dam supporters' vision of development is the direct outcome of a masculinist worldview (Kurian, 2000).

Supporters of the dam reject the participation of the displaced people in the decision-making process. This is what Kurian views as the representation of the masculinist worldview and the thwarting of environmental justice. And as a result, the costs borne by the most marginalized sections of society are not included in the government's calculations for how much compensation the entire state of Gujarat should receive. When compensation for the development of the dam is given under Gujarat's resettlement and rehabilitation (R&R) policy, it goes exclusively to the men. Single adult women who are either unmarried, widowed, or divorced are assumed to be part of a large family unit that has an adult male. Thus, norms of justice and equality are ignored for displaced people and specifically for women in these social groups (Kurian, 2000).

Those who oppose dam-building question the integrity of the development project. Kurian says that the opponents consider development as a modern concept that fosters inequity, destroys the environment, and serves as an ideological means of institutionalizing oppression. She points out that the people under the affiliation of the NBA she interviewed have differing ideological leanings. Some critics of the SSP advocate the use of appropriate technology; others question the social distribution of costs and benefits; and still others emphasize the need for an 'alternative political culture' along Gandhian lines, as Baviskar states (1995, p. 243).

Kurian argues that critics of the SSP who argue for appropriate technology demonstrate no awareness whatsoever of the fact that what is appropriate is gender-specific. By contrast, those who advocate the redistribution of social costs and benefits view such redistribution from an ecological Marxist perspective that places political and economic change above ecological concerns (Guha, 1988). Similarly, the Gandhian approach has not challenged the gendered nature of technology, although it is critical of modern technology (Kurian, 2000). Kurian concludes that the struggle in the Narmada Valley has been based on patriarchal power and contrasts the male-dominated vision of 'power over' with the feminist vision of empowerment understood as 'power to' (Kurian, 2000, p. 853).

3.3.1 Roles for Displaced Women

While the NBA was largely successful in securing rehabilitation for oustees, women were denied the right to land compensation because the Indian government perceived that only the men in a household should have ownership. Women were viewed as caregivers (Basu & Silliman, 2000). The movement's efforts successfully ensured that the lack of legally recognized titles to land in tribal societies did not stand in the way of tribes being compensated for the loss of land. As Misra (2007) notes,

> at the same time, however, women have been disempowered from land ownership rights. This has further strengthened the position of men in tribal communities because of the Indian government's particular conception of the household and the woman's position within it. (p. 137)

Medha Patkar, the leading proponent of the NBA movement, and other activists such as Arundhati Roy have also focused on the increase in tribal women's workloads in resettlement sites. They have highlighted the inaccessibility of fish from the Narmada River (the main source of protein in tribal diets) as a result of the government's decision to sell the rights to fishing in reservoirs to private parties (Misra, 2007). Roy, an author and activist of international renown, helped amplify the concern over women's rights both nationally and internationally in her book *The Greater Common Good* (1999). It was difficult to match the skills of women with the jobs available in the resettlement area (Basu & Silliman, 2000). Patkar and other female activists played a key role in the struggle by inspiring women to find work outside of their homes (Baviskar, 1995).

One might question whether gender justice can be achieved in resettlement and rehabilitation (R&R) policies in India. Mehta says that R&R programmes did not meet even the practical or material needs of women. If R&R can ever be seen as a 'development opportunity' (Mathur & Marsden, 1998) – and this in itself is questionable – it will need to be radically re-conceptualized. Such a re-conceptualization includes the need to avoid displacement or at least ensure that it is minimized. Mehta (2009) argues that, when displacement is unavoidable, it is important to:

- include women as full beneficiaries of compensation in their capacity as independent co-owners of land;
- make provisions for women's livelihood along with housing and employment;
- recognize and build on women's informal rights in customary practices;
- avoid any violation of their rights;
- include strong gender analysis and gender-sensitive data regarding the impacts of displacement;
- have special provisions to include the full participation of women in decision-making processes around displacement and resettlement;
- build strong safeguards to facilitate women's access to compensation and any other benefits. (pp. 28-29)

Another perspective on women and the NBA is found in Krishna's book *Genderscapes: Revisioning National Resource Management* (2009). Krishna examines three threads of environmentalism with regard to the struggle in Narmada Valley: global environmentalism, advocacy, and Vandana Shiva's version of ecofeminism. The global thread is exemplified by Indira Gandhi capturing the attention of the world at the first United Nations conference on the human environment at Stockholm in 1972 with her comment that 'poverty is the greatest polluter'. She proclaimed that it is through the instrument of development that the quality of life can be improved, and she managed to institutionalize environmental concerns in law and administration by setting up the National Committee on Environmental Planning and Coordination. Her political efforts resulted eventually in the establishment of the Ministry of Environment which in the late 1980s became the Ministry of Environment and Forests (Krishna, 2009).

With respect to advocacy, from the perspective of women, it is remarkable that the activist Patkar did little to address her own subordinate and oppressed position within her family. The same is true of many of the other women involved in the movement. Krishna (2009) points out that Patkar somewhat conservatively treats rural society as a homogenous whole, 'undifferentiated by class, caste, tribe and gender' (pp. 323-324).

The third thread examined by Krishna is Vandana Shiva's brand of back-to-nature ecofeminism. This type of ecofeminism relates women's procreative power with the feminine principle of nature and with women's role in subsistence production and forestry. Ecofeminists argue that female biology gives women a monopoly over holistic and ecological knowledge that men do not have. But Krishna says that in emphasizing women's biological, maternal, and procreative role, ecofeminism mirrors the very motifs of conventional development that it seeks to repudiate. Unless women's family and social status are strengthened, this kind of ecofeminism has only limited value to women (Krishna, 2009).

3.4 Gendered Dimensions of Neoliberal Capitalist Development

While the struggles of the Narmada communities received widespread attention, less noticed were the processes at work as part of a wider historical trend toward neoliberal capitalist development. The stakes are higher than the mere construction of a dam; it can be conceptualized as a struggle between the winners and losers of 'primitive accumulation', which concentrates property in a few hands while reducing the access of many to an independent means of livelihood. This process inevitably has gender dimensions, especially related to women's changing relation to property, both common and individual, caused by displacement and resettlement (Whitehead, 2003).

The marginality of the scheduled tribes in India stands in contrast to their predominance in the populations displaced by dams and other development projects. Although scheduled tribes (*adivasis*) make up 8 percent of India's population, they constitute between 40 to 50 percent of those whose lands have faced submergence in the post-independence period (Nayak, 2015). Census maps of Gujarat state, the site of the Sardar Sarovar Dam, show that of the 24 submerged areas of large dams built in that state after independence, 18 of them (or 75 percent) are located in predominantly tribal areas (Whitehead, 2003).

It is evident that large dams have widespread consequences for the economic, social, and cultural contexts within which men and women live their lives. Any analysis of large dams needs to focus on the extent to which they have changed or have the potential to change gender disparities, making it important to understand existing gender disparities in society and how they change in the course of the project's life (Adams, 2000; Mehta & Srinivasan, 2000). Another important point is that the men and women of the affected communities are not passive victims or recipients of dam-based development. Human beings are capable of tremendous resilience and can adjust to and act upon a wide range of changes. However, it has been demonstrated that dam-building activities have largely failed to make conscious efforts to consider equity considerations. Rarely have equity considerations been placed upfront in planned interventions associated with large dams. For development to be just, projects must aim to reduce inequality (both existing and new) by making the better off pay for most of the costs and allowing the worse off to receive most of the benefits (Adams 2000; Mehta & Srinivasan, 2000).

At the international level, there is a growing consensus about the need to achieve gender equality. The Fourth World Women's Conference in Beijing in 1995 emphasized the curtailment of gender inequality both as an end in itself and as something that can contribute to sustainable development. As Mehta and Srinivasan (2000) point out, bilateral and multilateral agencies are currently engaged in processes that seek to make gender equality mainstream in all areas of their work such as poverty, health, and environmentally sustainable development.

Notwithstanding these international commitments, noticeable gaps remain. Mehta and Srinivasan argue that the goal of gender equality in policy and practice should be unequivocally embraced by multilateral, international, and national dam-building agencies. The failure to do so would make the distribution of benefits and costs highly unequal in gender terms. Moreover, the lack of concern for equity could also undermine efficiency, which is a high priority in dam-building activities (Mehta & Srinivasan, 2000).

> An evaluation of impacts should hinge on two principal axes: equity and distribution. Gender scholars have long been concerned with issues

> concerning equity and distribution. Their primary concern has been to understand the root causes of gender gaps in the ways which the fruits and costs of development are borne. Thus, it has been necessary to understand discrepancies in the allocation and distribution of resources. (Adams, 2000; Mehta & Srinivasan, 2000, p. 6)

As Adams (2000) remarks: 'Thus costs and benefits have to be seen within a broader framework that acknowledges the heterogeneity of communities and the multi-layered nature of the social impacts of large dams' (p. 10). In accessing resources, equity and distribution are interconnected, and they apply to both men and women. Similarly, the costs and benefits of large dams would have to take into account the social, cultural, and political dimensions. Large dams have a lasting impact on kinship structures and community identities (Patel, 1997). Kinship structures are social resources that determine individual well-being and therefore cannot be ignored in impact assessment studies (Mehta & Srinivasan, 2000).

The cost-benefit analysis that was made to justify the building of dams was specifically designed to measure the costs or profits of projects such as dams. It is easy to calculate the economic costs or benefits. But how do cost-benefit analyses capture the socio-cultural identity and geographical spaces that are pivotal to the community's sense of well-being? Critics of conventional cost-benefit analyses argue that any social impact analysis must necessarily capture the hidden and invisible nature of changes and costs, which are not reflected in economic cost-benefit analyses (Kabeer, 1994; Elson, 1998; Mehta & Srinivasan, 2000).

3.5 Reasons for the Success of the NBA

Reddy argues that the NBA was the most popular movement in the environmental history of India because it focused on the issue of human rights. Medha Patkar and the other main leaders of the movement such as Baba Amte fought for proper rehabilitation programmes for those who were displaced by the construction of the dam. Reddy (1998) notes that 'Due to improper implementation of the rehabilitation programmes by the state of Gujarat, the human rights activists have become the articulators of anti-dam protests' (p. 688). They demanded a complete stop to the building of the dam as well as resettlement and rehabilitation benefits to the oustees (Wood, 1993). Environmentalists who opposed the construction of large dams for ecological reasons also supported the movement's demands (Reddy, 1998).

The movement received public attention, as it impacted three states. Indeed, as Reddy observes, 'the most notable feature of this movement is the international support it has received. In fact, the main reason behind the World Bank's withdrawal of funding for the project was due to international pressures' (Reddy, 1998, p. 688). For instance, Japanese environmentalists persuaded their government to block the money it had advanced for the project. Similarly, U.S. environmental groups have worked hard to put an end to World Bank funding (Gadgil & Guha, 1994). With this international support, the leaders of the movement were warmly received at a U.S. Congressional Sub-Committee hearing. In the wake of the World Bank's independent review, the European Parliament urged its member countries to instruct their World Bank directors to suspend all further aid to Sardar Sarovar, and in the final voting, all the donor countries voted against funding while poor recipient countries voted for continuation (Wood, 1993). This kind of international support had never before happened in Indian environmental history. The completion of the Narmada Valley Development Project (NVDP) is not directly linked with World Bank funding, but without international pressure, it would have been very difficult for the leaders of NBA to stop World Bank funding on their own. This may be viewed as only a partial success of the movement, as the ultimate measure of success would be convincing or pressuring the government to stop the project with people's support, which is most unlikely to occur given the present socioeconomic conditions in India (Reddy, 1998).

It is interesting to ask why the NBA became such a popular movement and why it received so much support from international environmental organizations. The NBA had multiple objectives. Like the Chipko movement, the NBA started with the issue of the survival of the indigenous people (the tribals), whose livelihood was being taken away. The second objective was to defend human rights. The third objective was to focus on the environmental concerns along with suggesting an alternative development paradigm. The strategy of the leaders was to address all three dimensions in an effective way, and they succeeded in doing so. According to Reddy, it is crucial to realize that the mass base of the movement was comprised of the dam-affected tribal communities, which are homogeneous and removed from modern influence and hence easier to organize. The main reason for its international recognition was the movement's timing. The NBA took off during the late 1980s, when the North-South divide regarding environmental issues was most pronounced. Protecting the environment in the Global South had become one of the main ideological objectives of the Global North. This contributed to the recognition and active support for the NBA from governmental and non-governmental organizations abroad. Reddy argues, 'International support was thus instrumental in bringing pressure on the World Bank to stop the funding of a project perceived in essence as anti-environment' (1998, p. 690).

3.6 Conclusion

In addition to founding the NBA, Medha Patkar was one of the founders of the National Alliance of People's Movements (NAPM), a non-electoral, secular political network of activists formed in 1996 that advocated an alternative development paradigm and opposed globalization and liberalization-based economic policy. As Prasad notes:

> Medha Patkar, founder of the NBA and the National Alliance of People's Movements, is a powerful voice for millions of voiceless poor and oppressed people whose uncompromising insistence on the right to life and livelihood has compelled the post-Independence generation in India as well as around the world to revisit the basic questions of natural resources, human rights, environment and development. (2013, p. 103)

NAPM has helped women to take part in a number of leadership roles not only as elected officials in local governments but also in tasks involving communicating with state officials and informing the communities and villages and raising their awareness. Although there is still a long way to go to convince the government to make the necessary policy changes related to development programmes, Prasad says that NAPM and other autonomous organizations in India are slowly making headway in exerting their power within the political process (2013).

This chapter has described how the Sardar Sarovar Project (SSP) was initiated immediately after India's independence as part of the mass industrialization vision of the first Indian prime minister, Jawaharlal Nehru. The history of dam-building – from its beginning in the colonial period through to its imitation by American engineers and its subsequent reintroduction in India which, taken together, constitute a full circle – is important for understanding how the protest movement started in the 1980s. Nehru inaugurated the SSP in the Narmada River, which travels through the three states of Gujarat, Maharashtra, and Madhya Pradesh. The chapter offers a detailed narrative of how all the authorities of the central and state governments became involved in the project and how the Narmada Bachao Andolan (NBA) emerged as an anti-SSP group. A pivotal moment in the history of the NBA is when it was able to convince the World Bank to withdraw its loan, after which the NBA took its case all the way to the Supreme Court of India. Indeed, the NBA has changed the economic development discourse not only in India but globally. The NBA is not only working on the rehabilitation of the most vulnerable displaced people in this region but has also opened up the discourse on the alternatives to dam-building which are more cost-effective – such as rainwater collection tanks and mini water projects that involve the local communities – compared to the high cost of building dams, which includes submergence of the local area, the rehabilitation and resettlement of the affected people, and the environmental impact (Prasad, 2013). Finally, in highlighting the chief reasons for the success of the NBA, this chapter has conducted an extensive survey of the literature to demonstrate that the NBA's impact in India and globally on the discourses on development, biodiversity, and sustainability has resulted in a paradigm shift towards including the fringe voices in the mainstream elite discourse (Prasad, 2013). This is a success story for India (Jebaraj, 2010; Prasad, 2013).

Another important point to note here is that after two decades of negotiation with the United Nations, India and seventeen other countries with rich natural resources that are being exploited were able to agree on an international treaty known as the Nagoya Protocol (adopted in 2010 and enforced from 2014) that allows for the sharing of benefits from the utilization of genetic resources with local communities. The Protocol is a plan to protect the biodiversity by creating legal certainty and transparency for both the providers and users of genetic resources. The access and benefit-sharing (ABS) clearing-house is the key tool to facilitate and implement the Nagoya Protocol, which requires multinational corporations to share its profits with the local communities not only for using the natural resources 'but also any derivative products developed from it' (Prasad, 2013, p. 106). For example, if a pharmaceutical company discovers a medicine from using plants in India, the company must pay not only the local community that has nurtured the natural resource but also for the permission to use the genetic material such as pathogens – whereas in the past the pharmaceutical companies used the pathogens, came up with vaccines, and made billions of dollars of profits for themselves without paying anyone. It is striking that out of the 200 countries that have signed the Nagoya Protocol, the United States is not one of the signatories.

In the next chapter, I turn to a contemporary environmental movement of India: Navdanya (Nine Seeds). This is a movement related to food sovereignty and, in particular, genetically modified organisms (GMOs). As will be shown, the Navdanya movement is built upon themes similar to those adhered to by the Chipko and NBA movements, namely nonviolence and feminism.

Works Cited

Adams, William (2000). 'The Social Impact of Large Dams: Equity and Distributional Issues'. Final Version (Report prepared for the World Commission on Dams Thematic Review: Social Impacts).

Agoramoorthy, Govindasamy, and Minna J Hsu (2008). 'Small Size, Big Potential: Check Dams for Sustainable Development'. *Environment Science and Policy for Sustainable Development*, 50, 22-35.

Basu. Pratyusha (2010). 'Scale, Place and Social Movements: Strategies of Resistance Along India's Narmada River'. *Revista NERA*, 13, 16, 96-113.

—, and Jael Silliman (2000). 'Green and Red, Not Saffron: Gender and the Politics of Resistance in Narmada Valley'. In Christopher Key Chapple and Mary Evelyn Tucker (eds.), *Hinduism and Ecology: The Intersection of Earth, Sky and Water.* Cambridge, MA: Harvard University Press, 423-452.

Baviskar, Amita (1995). *In the Belly of the River: Tribal Conflicts over Development in the Narmada Valley.* New Delhi: Oxford University Press.

Blewitt, John (2015). *Understanding Sustainable Development,* second edition. Milton Park: Routledge Taylor & Francis Group.

Boserup, Ester (1970). *Women's Role in Economic Development.* New York: St. Martin's Press.

Brooks, Karl Boyd (2006). *Public Power, Private Dams: The Hells Canyon High Dam Controversy.* Seattle: University of Washington Press.

Cullet, Philippe (2001). 'Human Rights and Displacement: The Indian Supreme Court Decision on Sardar Sarovar in International Perspective'. *The International and Comparative Law Quarterly,* 50, 4, 973-987.

Cernea, Michael M., and Scott Guggenheim (eds) (1993). *Approaches to Resettlement Policy: Practice and Theory.* Boulder, CO: Westview Press.

De, Debasree (2014). 'Nehruvian Vision of Sustainable Development for Tribals in India: A Critique'. *South Asia Research*, 34, 1, 1-18.

Dharmadhikary, Shripad (2005). *Unraveling Bhakra: Assessing the Temple of Resurgent India.*Badwani, India: Manthan Adhyayan Kendra.

Duara, Prasenjit (2015). *The Crisis of Global Modernity.* Cambridge: Cambridge University Press.

Dubash, Navroz K. et al. (2001). *A Watershed in Global Governance? An Independent Assessment of the World Commission on Dams.* Washington, D.C.: World Resources.

Elson, Diane (1998). 'Talking to the Boys: Gender and Economic Growth Models'. In Cecille Jackson and Ruth Pearson (eds.), *Feminist Visions of Development: Gender Analysis and Policy*. London: Routledge, 155-170.

Gadgil, Madhav, and Ramachandra Guha (1994). 'Ecological Conflicts and Environmental Movement of India'. *Development and Change*, 25, 101-136.

Gilmartin, D. (1995). 'Models of the Hydraulic Environment: Colonial Irrigation, State Power and Community'. In David Arnold and Ramachnadra Guha (eds.), *Nature, Culture, Imperialism: Essays on the Environmental History of South Asia*. New Delhi: Oxford University Press, 237-259.

Guha, Ramachandra (1988). 'Ideological Trends to Indian Environmentalism'. *Economic & Political Weekly*, 23, 49, 2578-2581.

International Rivers website: https://www.internationalrivers.org/resources/protesters-assemble-to-oppose-sardar-sarovar-dam-1968 (accessed 10 March 2018).

Jebaraj, Priscilla (2010). 'Nagoya Protocol a Big Victory for India says Jairam Ramesh'. https://www.thehindu.com/news/national/lsquoNagoya-Protocol-a-big-victory-for-India/article15798391.ece (accessed 16 July 2020).

Jensen, Robert (2004). 'Damn the Dams: An Interview with Medha Patkar'. *Alternet*, February 26.

Kabeer, Naila (1994). *Reversed Realities: Gender Hierarchies in Development Thought.* London: Verso.

Kingsbury, Benedict, and Doreen Lustig (2006). 'Displacement and Relocation from Protected Areas: International Law Perspectives on Rights, Risks and Resistance'. *Conservation and Society*, 4, 3, 404-418.

Krishna, Sumi (2009). *Genderscapes: Revisioning Natural Resource Management*. New Delhi: Zuuban.

Kurian, Priya A. (2000). 'Generating Power: Gender, Ethnicity and Empowerment in India's Narmada Valley'. *Ethnic and Racial Studies*, 23, 5, 842-856.

Leslie, Jacques (2005). *Deep Water: The Epic Struggle over Dams, Displaced People, and the Environment*. New York: Farrar, Straus & Giroux.

Manibeli Declaration. http://www.rivernet.org/manibeli.htm (accessed 28 May 2020).

Mathur, Hari Mohan, and David Marsden (eds) (1998). *Development Projects and Impoverishment Risks: Resettling Project-Affected People in India*. New Delhi: Oxford University Press.

McCully, Patrick (2001). 'The Use of a Trilateral Network: An Activist's Perspective on the Formation of the World Commission on Dams'. *American University International Law Review*, 1453-1475.

Mehta, Lyla (2008). 'Why are Human Rights Violated with Impunity? Forced Displacement in India's Narmada Valley'. In Katarzyna Grabska and Lyla Mehta (eds.), *Forced Displacement: Why Rights Matter*, UK: Palgrave Macmillan, chapter 10. 201-221.

— (2009). 'The Double Bind: A Gender Analysis of Forced Displacement and Resettlement'. In Lyla Mehta (ed.), *Displaced by Development: Confronting Marginalization and Gender Injustice*. New Delhi: Sage Publications, chapter 1, 3-33.

—, and Bina Srinivasan (2000). 'Balancing Pains and Gains: A Perspective Paper on Gender and Large Dams' (Report prepared for the World Commission on Dams Thematic Review: Social Impacts).

Misra, Shalini (2007). 'Spirituality, Culture and the Politics of Environmentalism in India'. *Journal of Entrepreneurship*, 16, 131-145.

Morse, Bradford, and Thomas R. Berger (1992). 'Sardar Sarovar – Report of the Independent Review', 349-358. http://www.ielrc.org/content/c9202.pdf (accessed 20 January 2018).

Narula, Smita (2008). 'The Story of Narmada Bachao Andolan: Human Rights in the Global Economy and the Struggle Against the World Bank'. *The New York University Public Law and Legal Theory* Working Papers, Paper 106, 351-383.

Nayak, Arun Kumar (2015). 'Environmental Movements in India'. *Journal of Developing Societies*, 31, 249-280.

Parasuraman, S. (1993). 'Impact of Displacement by Development Projects on Women in India'. *ISS Working Paper Series*, 159. The Hague: Institute of Social Studies of Erasmus University Rotterdam (ISS).

Patel, Anil (1997). 'What do the Narmada Valley Tribals Want'? In W.F. Fisher (ed.), *Toward Sustainable Development: Struggling Over India's Narmada River*. Jaipur: Rawat Publications, 179-200.

— (1997). 'Resettlement Politics and Tribal Interests'. In Jean Dreze, Meera Samson, and Satyajit Singh, (eds.), *The Dam and the Nation: Displacement and Resettlement in the Narmada Valley*. New Delhi: Oxford University Press, 66-92.

Penz, Peter, Jay Drydyk, and Pablo S. Bose (2011). *Displacement by Development: Ethics, Rights & Responsibilities*. Cambridge: Cambridge University Press.

Peterson, M.J. (2010). 'Narmada Dams Controversy'. *International Dimension of Ethics Education in Science and Engineering* www.umass.edu/sts/ethics/narmada.html (accessed 15 February 2020).

Prasad, Kiran (2013). 'Environmental Communication from the Fringes to Mainstream: Creating a Paradigm Shift in Sustainable Development'. In Jan Servaes (ed.), *Sustainable Development and Green Communication: African and Asian Perspectives*. UK: Palgrave Macmillan, 95-109.

Rajan, S. Ravi (2005). 'Science, State and Violence: An Indian Critique Reconsidered'. *Science as Culture*, 14, 3, 10-12.

Ramachandra, Komala (2006). 'Sardar Sarovar: An Experience Retained'. *Harvard Human Rights Journal*, 19, 275-81.

Rangachari, R. (2006). *The Bhakra-Nangal Project: Socio-Economic and Environmental Impacts*. New York: Oxford University Press.

Reddy, V. Ratna (1998). 'Environmental Movements in India: Some Reflections'. *Journal of Indian School of Political Economy*, 10, 4, 685-95.

Roy, Arundhati (1999). *The Greater Common Good.* Mumbai: India Book Distributor.

— (1999). *The Cost of Living.* New York: The Modern Library.

Scudder, T. (1995). 'Resettlement'. In A. Biswas (ed.), *Water Resources: Environmental Planning, Management, and Development.* New York: McGraw Hill.

Supreme Court of India Judgement (2000). WRIT PETITION (C) NO. 319 oF 1994: *Judgement,* New Delhi. At: www.narmada.org/sardarsarovar/sc.ruling/majority.judgement.htm (accessed 9 February 2018).

Thukral, Enakshi Ganguly (1996). 'Development, Displacement and Rehabilitation: Locating Gender'. *Economic and Political Weekly*, 31, 24, 1500-1503.

Whitcombe, Elizabeth (1972). *Agrarian Conditions in Northern India: The United Provinces under British Rule.* Berkeley, CA: University of California Press.

Whitehead, Judy (2003). 'Space, Place and Primitive Accumulation in Narmada Valley and Beyond'. *Economic and Political Weekly*, 38, 40, 4224-4230.

Wood, John R. (1993). 'India's Narmada River Dams: Sardar Sarovar under Siege'. *Asian Survey*, 33, 10, 968-984.

World Commission on Dams, Dams and Development: A New Framework for Decision Making (English) (2000). https://www.internationalrivers.org/sites/default/files/attachedfiles/ world_commission_on_dams_final_report.pdf (accessed 10 April 2018).

World Commission on Environment and Development (1987). *Our Common Future.* Oxford: Oxford University Press (The Brundtland Commission Report).

Worster, Donald (1985). *Rivers of Empire: Water, Aridity, and the Growth of the American West.* New York: Pantheon.

4 Navdanya (Nine Seeds) Movement

Abstract

This chapter addresses the issue of sustainable agriculture by relating it to food sovereignty and food security and the contribution that Navdanya is making by promoting *seed satyagraha,* or civil disobedience against seed control by large corporations. The founder of Navdanya, Vandana Shiva, is highly critical of GM crops and chemically based industrial agriculture, which are monopolized by multinational companies in India. Shiva's philosophy of Earth Democracy is based on the premise of the interconnectedness of everything. Agroecology must be transformed based on sustainable agriculture, which is possible only by promoting local farmers who use organic seeds from community seed banks such as those started by Navdanya and can protect their livelihood and communities as well as the natural resources.

Keywords: GMOs (genetically modified organisms), Earth Democracy, Vandana Shiva, food sovereignty, sustainable agriculture, La Via Campesina

In the 1960s and 1970s, the Green Revolution in India provided food security for a nation whose population had more than doubled since independence in 1947. Agricultural productivity was achieved through heavy reliance on fertilizers and pesticides as well a few high-yield seed varieties, sown in the agricultural practice of monocropping – growing a single crop repeatedly on the same land to increase efficiency – which in turn reduces agricultural diversity. Decades of monocropping have led to soil salinization and degradation, along with major increases in water consumption through public subsidies and abusive practices (Vyas, 2003) as well as accelerated groundwater depletion even in such rain-fed states as Andhra Pradesh, Karnataka, Rajasthan, Madhya Pradesh, Chhattisgarh, and Gujarat (Virmani & Lepineux, 2015).

The Navdanya movement is a struggle for freedom against large transnational corporations and their partnered states for the most fundamental

Mallick, Krishna, *Environmental Movements of India: Chipko, Narmada Bachao Andolan, Navdanya.* Taylor & Francis Group, 2021.
DOI: 10.5117/9789462984431_CH04

survival needs: food and water. In this movement, independent and indigenous farmers practiced civil disobedience by refusing to follow laws constructed and enforced by the very states charged with protecting their fundamental human rights (Armaline & Glasberg, 2009). Agricultural biotechnology is at the forefront of the international development discourse. The debate around genetically modified (GM) crops has widened to include the future of agriculture and small-scale farmers, corporate control, property rights, and the rules of global trade (Scoones, 2008). Some argue that one of the most efficient ways to feed the world, especially in developing countries with increasing populations, is to cultivate GM crops. Others claim that the high cost of GM crops and the lack of diversity with respect to seed choice lead to domination by corporate-capitalist agriculture.

India's exclusive focus on food security in the 1960s and 1970s was understandable given its need to secure a minimum quantity of available food for India's 1.3 billion inhabitants. From the 1980s onwards, however, food security raised broader questions about the environment, sustainability, and farmers' livelihood. Environmental movements that defended small farmers and producers against large-scale development programmes drew the attention of small farmers who had given up farming. Virmani and Lepineux point out that according to census data between 1991 and 2001, nearly eight million people abandoned farming (p. 126).

4.1 The Terrible Human Toll of GM Crops

India's response to monocropping and GM crops has been much more tragic than a loss of livelihood. Since 1990, an estimated 300,000 farmer suicides have taken place in the cotton-growing districts of central and southern India (Philpott, 2015; Mishra, 2014). The devastating loss of life has been alternatively attributed to increased indebtedness (Gruere et al., 2008), the introduction of *Bacilus thuringiensis* (Bt) cotton, cash cropping (crops grown for direct sale in the market), industrial agriculture, and globalization (Mahadevan, 2003). Other factors that may have driven farmers to despair were increased production costs, the fear of crop failure, and developed-country subsidies that lowered world prices for cotton.

Those who oppose biotechnology claim that the driving force behind the farmer suicides is Monsanto's monopolization of the cotton seed sector. Monopolization involves the control of intellectual property as well as the restriction of choices. Monsanto's GM seed, the only lab-altered crop in India, was introduced in 2003 and upgraded in 2006. While helping India become the world's top producer and second-largest exporter of the fiber, Monsanto acquired 90 percent of India's cotton acreage.

Back in 1990, Monsanto had requested the authorities in India to conduct field trials, but in 1993, this request was rejected by India's Department of Biotechnology (DBT). The DBT is the absolute authority for monitoring elementary and small-scale tests, due to the high trait fees and the problems that might arise as a result of crossbreeding an American variety with an indigenous one (Gupta & Chandak, 2005). Yet in a turnaround five years later, Mahyco was allowed to conduct field trials with local varieties crossbred with Monsanto's imported 100g gene. In that same year, Monsanto bought a share of Mahyco and participated directly in the experiment. This resulted in the creation of Mahyco Monsanto Biotech (MMB) (Gupta & Chandak, 2005; Newell, 2007; Thomas & Tavernier, 2017).

Due to allegations of monopolization, MMB's Bt cotton was banned in August 2012 in Maharashtra, but this ban was subsequently lifted in May 2013 (Thomas & Tavernier, 2017). According to Virmani and Lepineux (2015), 'The needs of India's growing population and the unstable trends of agricultural production, coupled with the outburst of farmer suicides, gave urgency to the need to reform the mechanisms and regulatory institutions of rural farming' (p. 126).

The primary method of differentiating GM crops from hybrid varieties has been through 'terminator technology' frame. Terminator technology was originally developed as a 'technology protection system', part of a class of genetic use restriction technologies, and was intended to be a way of stopping the re-sowing of seeds by producing sterile ones. The technology was initially developed by Delta and Pine Land in 1998, which was subsequently acquired (along with its patents) by Monsanto in 2006.

Vandana Shiva, the founder of Navdanya, accuses MMB of criminalizing the farmers' practice of seed saving by manipulating the World Trade Organization's Agreement on Agriculture by way of the sections on Intellectual Property Rights. Thomas and Tavernier say that in Shiva's opinion, Article 27.3 of the Trade-Related Intellectual Property Rights' (TRIP) agreement was incorporated into the Indian Patent Act to prevent farmers from seed saving and seed exchange, and as a result their livelihood was threatened (2017). However, Herring (2008) shows that when Shiva first started criticizing Monsanto for seed monopolization, Monsanto was still grappling with field trials. The proposed 2004 Seed Bill, which forbids farmers from selling unregistered seeds, has yet to be passed by the Indian parliament because of the public outcry against it (Peschard, 2014). As Thomas and Tavernier point out, the MMB seeds are not patent-protected like the seeds of other firms, and therefore the proponents of biotechnology argue that the rising rate of farmer suicides cannot be attributed solely to MMB's patent monopoly. They claim that MMB has benefited not from a monopoly but from an advantageous market position due to India's bio-safety regulations (Herring, 2015; Thomas & Tavernier, 2017).

4.1.1 Shifting Economics

Before the introduction of GM crops, 50 percent of India's national income had come from agriculture and allied activities, employing 72 percent of its working population (Virmani and Lepineux, 2015). This raised questions about the type of national development strategies that promote sustained agricultural growth in a rural landscape marked by a diversity of climate and crop patterns and practices (Ninan & Chandrashekar, 1993). In particular, the expansion of sown areas became an important subject for debate (Tripathi & Prasad, 2009). The idea that reductions in yield could be linked to land and water use and exploitation began to gain ground, sustainable agriculture became a major concern in India. By the end of the 20th century, agricultural and development policies received widespread national and international media coverage that generated a barrage of reports and debates (Krishnakumar, 2005; Sahai et al., 2005; Nadal, 2007; Virmani & Lepineux, 2015).

Virmani and Lepineux point out that the main challenges for Indian agriculture were now extended to minimizing the environmental impacts and land degradation and preserving resources (land, water, air). India's National Agricultural Policy (NAP) placed sustainable agriculture and the importance of the management and conservation of resources at the heart of its focus. 'The policy seeks to promote technically sound, economically viable, environmentally non-degrading, and socially acceptable use of country's natural resources for sustainable development of agriculture' (Government of India, 2000). Its highest priority was to improve the quality of the land and the soil by making the farming community aware of the environmental concerns and by promoting more rational utilization and conservation of water. The tenth Five Year Plan for 2002-2007 emphasized natural resource management through rainwater harvesting and controlled groundwater exploitation. It did not allow the abusive use of fertilizers and recognized organic farming as a 'thrust area' (an area of focus) in the sustainable use and management of resources in agriculture (Government of India, 2002; Virmani & Lepineux, 2015).

4.2 Emergence of Anti-GM Movements

As with all social movements, the protests against GM crops in India were and are heterogeneous and fluid, made up of groups and individuals with varying perspectives and levels of involvement. While most well-known opposition groups including Navdanya are based in the north, some of the largest mobilizations against GM crops have taken place in Karnataka in the south. It is important to note that while Navdanya and other movements were not coordinated, each in its own way has protested against genetically modified organisms (GMOs) in India (Stone, 2004).

4.2.1 KRRS: Fiery Fields of Protest

After Monsanto imported the Bt cotton gene in 1998, indigenous non-governmental organizations (NGOs) and activists began to protest against it. The *Karnataka Rajya Raitha Sangha* (KRRS), or State Farmer's Association, had during the 1980s primarily been a protest movement focusing on a host of agricultural issues, including terms of trade that were unfavourable to agriculture, declining per capita income, the purchasing power of farmers, increased costs of input prices, low prices for agricultural crops, the growing poverty rate, mounting debt, and urban-biased state policy (Assadi, 2004).

Changing tactics, in 1998 the KRRS joined the 'Cremate Monsanto' campaign, also known as 'Monsanto, Quit India'. The campaign was launched on 9 August by a coalition of NGOs, and the KRRS began its part of the campaign in November (Nanjundaswamy, 1998a). The KRRS warned that all trial fields in Karnataka would be burned in the presence of the media. The farmers of Bellary and Raichur districts consented to the arson. The crop-burning was repeated in 2002, following the official release of GM seeds. This protest had a similar outcome each time; some farmers received compensation from the KRRS while others resorted to police protection (Scoones, 2008). Apart from this media drama, the KRRS also demanded a five-year ban on GM cotton seeds (Scoones, 2005; Thomas & Tavernier, 2017).

The KRRS was instrumental in mobilizing farmers around Bt cotton, which they opposed as part of a broader resistance against neoliberal globalization. Neoliberal globalization leads to people throughout the world being affected by interrelated processes, and it also constrains the role of the state due to the increasing power of international financial institutions and multinational corporations (Croeser, 2015). While positioning its opposition to GM crops within the context of the Indian independence movement, the Cremate Monsanto campaign also situated it within the contemporary movement to resist neoliberal globalization. At the same time, there has been a growing movement within India to promote organic and traditional agriculture, much of which overlaps significantly with the anti-GM movement. Through activism inside India as well as international connections, this movement has been working to redefine the debate over GM crops.

Common to the framing of GM crops within both the Gandhian discourse and the opposition to neoliberalism is the positioning of GM crops as the extension of an existing trend, the commercialization of seed in the form of hybrids. In both frames of resistance, commercial seeds are seen as part of a shift away from local sovereignty. The anti-neoliberalism discourse emphasizes the specifically U.S. origin of GM seeds, objecting not just to the fact that seeds are genetically modified but also to the entry of foreign (particularly U.S.) firms into the Indian seed market. Similarly, within the Gandhian notion of *swadeshi* (honouring the local), there is significant opposition to buying seeds – genetically modified or otherwise – from outside the village.

4.2.2 Gene Campaign: Secure Food and Climate

Founded by Dr. Suman Sahai in 1993, the Gene Campaign is a grassroots-level research and advocacy organization committed to achieving food security and sovereignty. The organization's mission is to work for a just and equitable policy framework and to enable sustainable agriculture, self-reliant farming, and food for all. The focus of Gene Campaign's work is to ensure food nutrition, and a secure livelihood for rural and tribal communities. To achieve this, the organization works to conserve the genetic diversity of important crop plants, making small-farm agriculture sustainable and adapting it to climate change. It also attempts to fight malnutrition by increasing dietary diversity, using underutilized crops, and reviving forgotten traditional cereals, fruits, and vegetables.

Gene Campaign was the driving force behind a national campaign against seed patents and has worked with the government to construct a law on farmers' rights – the only country in the world to have done so. It continues to be active in the field of intellectual property rights (IPRs) and the legal protection of indigenous knowledge to ensure that the rights of rural and tribal communities are protected. Gene Campaign actively advocates laws and policies related to biodiversity, research, education, food, agriculture, and livelihood. Its field research in dry land areas such as in the states of Jharkhand and Uttarakhand provides a model for improving food production and shielding farmers against the worst impacts of global warming and climate change.

4.2.3 Navdanya's Holistic Approach

Navdanya promotes domestic crop biodiversity through its work with small farmers, primarily in the north of India (Navdanya, 2008a). It is a participatory research initiative founded in 1987 by Dr. Vandana Shiva as a programme of the Research Foundation for Science, Technology and Ecology (RFSTE). Active in 16 states of India, Navdanya mainly focuses on Uttarakhand in north India. It has trained 200,000 farmers in organic production and provides short-term training programmes as well as follow-up training sessions for farmers who are interested in being self-reliant using locally available resources. Hrynkow (2018) notes that this NGO has 100 community seed banks now spread across India, with model farms, cafés, and retail outlets in both New Delhi and Mumbai. Navdanya also conserves indigenous varieties of crops in organic farms spread over a 20-acre area in Uttarakhand.

The longest-held opposition against GM crops has come from Shiva and the associations she has founded – RFSTE and Navdanya – through a number of activist writings, public interest litigations (PILs), and agitation events such as seed tribunals attended by farmers from Europe, the United States, and India (Scoones, 2003). Thomas and Tavernier (2017) point out that Dr. Shiva's advocacy work began with a PIL in the Supreme Court of India on the integrity of the genetic engineering regulation procedures. She challenged the Review Committee on Genetic Manipulations (RCGM) that permitted the field trial by Mahyco-Monsanto, arguing that it had violated the 1989 Amendment of the Biosafety Rules, which vested the Genetic Engineering Approval Committee (GEAC) of the environment ministry and not the RCGM with the authority to permit trials.

Navdanya's main aim is to help bring about the seed conservation and seed exchange of traditional varieties. It is concerned with the conservation of biodiversity to maintain the cultural and material sustenance of people in India and outside. Navdanya's biodiversity conservation programme is aimed at supporting local farmers, rescuing and conserving crops and plants that are being pushed to extinction, and making endangered crops available through direct marketing. Navdanya is actively involved in the revival of indigenous knowledge and culture. It has created awareness of the hazards of genetic engineering, defended people's knowledge from biopiracy, and promoted food rights in the face of globalization (Kumari & Mallick, 2015). These goals are now being threatened by the international trade regime, as intellectual property rights being formulated by the World Trade Organization (WTO) are undermining farmers' rights to innovation and are encouraging the development of monopolies.

Navdanya calls international organizations such as the WTO 'the new colonizers'. Indeed, the protection of biodiversity is also part of the anti-globalization movements. Navdanya promotes products based on biodiversity by managing programmes such as Navdanya Foods, by organizing festivals and campaigns, and by distributing informative booklets and other political literature (Artner, 2004).

4.3 Food Sovereignty

Navdanya is unique because it focuses on the broader concept of food sovereignty. As Patel (2009a) explains, food sovereignty is the right of people to define its food and agriculture:

> It is a right, in other words, to have a say about the food system. It is, to use the words of the German political philosopher Hannah Arendt, a call for a right to have rights about the food system. It's a call that comes from those who have systematically been excluded from the formulation of food policy, who have long been forced to live with the consequences of agrarian policy authored by those in cities with few, if any, links of accountability to those whose lives are wrecked by their ideas.

In recent years, the concept of 'food sovereignty' has gained ground among grassroots groups and has taken the form of a global movement. As Monshipouri, Welch, and Kennedy (2003) explain, the global economy and forces of globalization have become distinguishing characteristics of the current paradigm of world politics, an arena that rests on the balancing claims and criticisms of multinational, or transnational, corporations (MNCs or TNCs). MNCs are given centre stage because of their dynamic growth and influence as well as the ways in which they affect the life chances of millions of people around the world. Yet certain corporate behaviour is detrimental to internationally recognized norms of human rights.

The idea of food sovereignty was introduced at the World Food Summit in 1996 by Via Campesina – the namesake of the International Peasant Movement – to address the fact that food security did not lead to a discussion of the social control of the food system. As a result, it is possible to have food security in prison or under a dictatorship. What mattered with respect to food security was merely food efficiency. Via Campesina argued that food sovereignty is a logical precondition for the existence of food security and defined it as follows:

> Long-term food security depends on those who produce food and care for the natural environment. As the stewards of food producing resources, we hold the following principles as the necessary foundation for achieving food security. Food is a basic human right. This right can only be realized in a system where food sovereignty is guaranteed. Food sovereignty is the right of each nation to maintain and develop its own capacity to produce its basic foods respecting cultural and productive diversity. We have the right to produce our own food in our own territory. Food sovereignty is a precondition to genuine food security. (cited in Patel, 2009b, p. 665)

Navdanya, on the other hand, defines food sovereignty as 'the right and freedom to grow diverse and nutritious food and the right to have access to safe, healthy, adequate and affordable food. Food sovereignty grows from household, to the community, the regional and the national level' (Navdanya.org). In the context of food sovereignty, it is important to address what Navdanya as a grassroots movement is doing and how it deals with the human rights issue by using the seed-procurement strategy. More precisely, the sections below will analyze the way in which Navdanya employs the Gandhian strategy of *satyagraha* within the complexity of the political framework of India and how it criticizes the profit-motivated activity of MNCs and the state.

The Nyéléni Forum documents that came out of the first Global Forum on Food Sovereignty in 2007 articulate what food sovereignty means, especially with regard to local control of resources and decision-making.

The notion of food sovereignty seeks to redefine relationships regarding markets, property, and governance. Nyéléni delegates determined that 'currently trade is based on an unsustainable production system and is controlled by [transnational corporations]', and they recommended returning democratic control of food distribution to producers and consumers as well as implementing 'autonomous control over local markets' (Nyéléni Declaration, 2007, pp. 25-26). They also sought to ensure inclusive and democratic control of all resources of production including water, land, and seed. They conclude with a simple statement: 'We will fight privatization and patenting' (Nyéléni Declaration, 2007, p. 35). In India, the fight against patents is framed as a *satyagraha,* or civil resistance, by the most ardent anti-GM activists.

4.3.1 Biodiverse Organic Farming

Sustainable agriculture has become a priority in India's development schemes over the last few decades within the government and among the NGOs who act in defence of civil society. Nowadays, organic farming is seen as contributing to the protection of natural resources, biodiversity, animal welfare, and the development of rural areas. With organic farming, the participating farmers are able to achieve economic stability and are given the opportunity to cultivate a wide range of produce. India currently exports organic products to Europe, the United States, and Japan.

The Government of India launched the National Programme on Organic Production (NPOP) in 2000 to promote further development in this area. Virmani and Lepineux (2015) posit that Navdanya's multi-pronged fight for farmers' rights and the common good, articulated in a language rooted in the traditional wisdom and knowledge of the common people, fits into this context. Its protest against WTO rules, seed patents, and the invasive strategies of multinationals is strengthened by its counterproposals for sustainable agriculture. Nevertheless, the shift from input-intensive farming to organic farming requires changes in a number of policy areas, from the transfer of information to the training and development of markets. Navdanya defines itself as a facilitator of these far-reaching changes.

The growing interest in organic agriculture alongside new entrepreneurial patterns encouraged a model that highlights a holistic, spiritual approach to farming. It was in this context of renewal, amid international support for alternative approaches to agriculture, that Navdanya was born in 1987 and continues to flourish. Its 25th anniversary was celebrated at the 2012 *Bhoomi* Festival of sustainable agriculture, seed sovereignty, and womanhood (Virmani & Lepineux, 2015). Navdanya promotes nonviolent farming, which protects biodiversity, the earth, and small farmers. Navdanya means 'nine crops or seeds', representing India's collective source of food security. The words *nav* (new) and *daan* (gift, offering), combined to make a compound word, signify renewal and rebirth of the principle that structures relations within the community, with one another, and with the gods (Virmani & Lepineux, 2015). Navdanya's mission focuses on improving the well-being of small and marginalized rural producers through nonviolent, biodiverse organic farming and fair trade. Biodiverse organic farming produces more food and nutrition and brings higher incomes to farmers than monocultures and chemical farming do.

4.4 Biodiversity and Climate Change

While avoiding environmental harm, biodiverse organic farming will also ensure successful crops in times of climate change. For Navdanya, 'climate change is a massive societal challenge which is not only converting, but also threatening and destroying the livelihood of the majority of the 1.2 billion inhabitants of the Indian subcontinent' (Harms & Powalla, 2014, p. 185). In addition to its concern for sustainable organic biodiversity and its mission to eradicate the poverty of small farmers, Navdanya is meeting the challenges of climate change through education. Bija Vidyapeeth (translated as 'Earth University') trains peasants to increase their harvest and prepares them for unstable weather. One of the areas that Navdanya works in is the Himalayas, where studies have shown an increase in temperature of about 0.6 degrees over the last 30 years, along with 10 to 60 metres (32.8 to 196.8 feet) of glacier melt (Harms & Powalla, 2014 , p. 185). Most of the population in northern India receives its water from this region. As Shiva and Bhatt (2009) warn, when the glaciers disappear and stop regulating the water supply in North India, there is a risk of total dryness in times of drought. Harms and Powalla (2014) have documented evidence of the consequences of climate change for peasants in the region between Ladakh and Dehradun. In 2009, after a severe drought leading to losses in their harvest, high amounts of precipitation followed, with dangerous cloudbursts. In 2010, dozens of people died due to a spring tide in Ladakh. In the following year, the monsoon was normal. For Navdanya and its members, therefore, climate change is a major source of insecurity.

The climate knowledge of Navdanya is integrated with its scientific work. Harms and Powalla point out that for Navdanya members, climate change has an effect on their everyday lives and is not based on scientific theory only. As Shiva (2016) notes, 'Navdanya wanted to build a program in which farmers and scientists relate horizontally rather than vertically, in which conservation of biodiversity and production of food go hand in hand, and in which farmers' knowledge is strengthened, not robbed' (p. 82). Another hurdle in this effort is the overconsumption of resources which is a violation of the duty that human beings have to respect nature at any cost. This duty has intrinsic and extrinsic value; any damage done to the biodiversity of nature affects the symbiotic balance and harmony.

In 2008 and 2009, Navdanya conducted a climate change campaign in Garwhal, a region at the foot of the Himalayas (Harms and Powalla, 2014). The campaign had two goals: to research climate change and inform residents about their findings. The campaign lasted for six days in which

Navdanya interviewed peasants in the region about their environmental experiences and distributed thousands of leaflets to the residents. This was followed up with workshops and conferences with the residents participating. This lasted for a few months, at the end of which 'veteran intellectuals' were selected from the peasants whom they had interviewed (Rawat, 2010). They were acknowledged as rural people who gave their expertise for a common plan to mitigate and adapt to climate change. After a few meetings, a common document was presented to the Minister for Environment and Forests of Uttarakhand in September 2009 (Harms & Powalla, 2014).

Due to Vandana Shiva's charismatic leadership and Navdanya's work with the local communities in India in an independent way through their actions and initiatives, it does not constitute the institutional arm of peasants' fights. Sometimes it might participate in international networks or collaborate with other activists, yet Navdanya's involvement with climate change takes place without a permanent political partnership. Shiva and Bhatt (2009) argue,

> Increasing the biodiversity of farming systems can reduce vulnerability to drought. Millets, which are far more nutritious than rice and wheat, use only 200-300 mm water, compared to 2500 mm needed for Green Revolution rice farming. However, global trade is pushing agriculture to GMO monocultures of corn, soya, canola and cotton, worsening climate vulnerability. (p. xii)

Harms and Powalla (2014) note that instead of trying to influence international law or change the national strategy, Shiva works at the grassroots level of activism. They refer to the view of environmental scientist and activist Sharachandra Lele, who points out that there are three broad currents of environmental movements in India – wildlife conservationists, livelihood movements and urban environmentalists. These three movements should be integrated under two conditions, that governments complement the multidimensional analysis of the roots and consequences of climate change with the linking of various forms of action. Harms and Powalla argue for the integration of Lele's two conditions. They point out that although South Asia is one of the regions in Asia most affected by climate change, the Indian government is still not critically engaged in this discourse. The identification of the root causes of climate change and the actions that need to be taken have yet to be integrated (Harms & Powalla, 2014).

4.4.1 Entrepreneurial Renewal

Navdanya is both a network of organic farmers and an entrepreneurial venture. Virmani and Lepineux trace its emergence in the context of transforming entrepreneurship patterns in India, which has changed significantly since the beginning of the 20th century. During its long history under colonial rule, India had entrepreneurs who were typically general traders, indigenous bankers, mill managers, and technicians. After independence, a system of permits and regulations for setting up businesses – known pejoratively as the License Raj – exercised control over industrial activities through government permits and the state regulation of markets. This stymied entrepreneurship within family businesses and put an end to the traditional paternalistic approach. Government control of production and assistance in the form of funds and infrastructural facilities further reduced entrepreneurial activities. Having the right contacts in government became important. The Nehruvian approach of the 1950s and 1960s perceived private enterprise as primarily benefiting individuals but also contradicted this by implementing large-scale state development plans for the nation. As a result, until the 1990s, entrepreneurship and business leadership were predominantly developed and transmitted within family firms or in well-entrenched large companies' trading communites. The end of the protectionist era in the 1990s marked a rupture with this traditional pattern of entrepreneurship (Virmani & Lepieux, 2015). A new model of entrepreneurship gained ground, once again made up of people from diverse social groups, classes, or occupational assemblies.

The climate of entrepreneurship in India is considered more favourable today than it was half a century ago because there is less state intervention, better managerial skills, and more focus on performance and turnover. Navdanya buys organic products from farmers with a 10 percent price premium. Most of the organic farmers are linked to the local market. In some instances, there is a direct marketing link between Navdanya's consumer members and producer members, and in a few cases, a door-to-door organic vegetable basket delivery is operated. Navdanya runs Earth University to 'spread the ideas and philosophies related to living seeds, living soil, living food, living economies and living democracies' (Shiva 2016c, p. 149; Hrynkow 2018, p. 206).

In a stark warning, Shiva (2014) proclaims that

> Today, the freedom of nature and culture to evolve is under violent and direct threat. The threat to seed freedom impacts the very fabric of

> human life and the life of the planet. Seed keepers, farmers and citizens around the world have joined together as a Global Citizens Alliance for Seed Freedom to respond to this Seed Emergency and to strengthen the movement for the freedom of humanity. (p. 219)

Navdanya is most visible in the Indian organic market, but it is not the only major organic player in India. Navdanya's distinguishing approach is to place marketing in a broader intellectual context by showing the flaws of the current system in India and encouraging qualitative changes in habits of production and consumption. But as we will see below, Navdanya is also profoundly interested in protecting human rights.

4.5 Navdanya and Social Justice

Vandana Shiva has been an active voice in the environmental movement for several decades, as Catherine Etmanski (2018) notes in her article 'Seeds and Stories of Transformation from the Individual to the Collective'. Shiva's work has focused on the destructive effects of monocropping, biotechnology, and genetically modified seeds on the Indian subcontinent in particular and on indigenous peoples worldwide. She argues that such practices were established primarily from India's participation in the Green Revolution and membership in the World Trade Organization as well as other free trade agreements and policies associated with globalization (Shiva, 2010; Shiva & Jalees, 2006). Shiva has dedicated her life's work to transforming global food systems and promoting an earth-centric approach to democracy that respects all beings, human and nonhuman alike (Etmanski, 2018).[2]

In her work, Etmanski points out that Shiva makes a direct connection between food sovereignty and social justice by following Gandhi's four pillars of nonviolence: *swaraj* (personal mastery, self-discipline, or localized self-governance); *sarvodaya* (the well-being of all creatures); *swadeshi* (honouring the local); and *satyagraha* (nonviolent noncooperation with unjust rules; for example, conscientious objection). As part of Navdanya's philosophy of *bij swaraj* (translated as seed sovereignty or local self-governance of seed), Shiva promotes the saving of seeds as a form of *seed satyagraha,* that is, civil disobedience against seed control by large corporations. This is similar to Gandhi's 1930 salt *satyagraha*, or Salt Marches against the British salt

2 For more information on Navdanya's campaigns and other works, see the Navdanya website, especially the section on Earth Democracy.

tax. Today, the multinational corporations represent the imperial power by controlling seeds rather than by occupying a nation (Etmanski, 2018). Shiva has described the direct relation between Gandhian principles and her current focus on seeds as follows:

> I thought of Gandhi's Spinning Wheel which had become such an important symbol of freedom, not because it was big and powerful, but because it was small and could become alive as a sign of resistance and creativity in the smallest of huts and poorest of families. In smallness lay its power. The seed too is small. It embodies diversity... In the seed, ecological issues could combine with social justice. (Navdanya, 2007, p. 8)

Etmanski (2018) mirrors this analogy with the spinning wheel when she writes:

> just as spinning *khadi* (home-spun cotton) became a symbolic way for everyday householders to participate in resistance against the dominance of the British textile industry, so too has saving seeds become a symbol for average people around the globe to participate in the resistance against the dominant industrial agri-food system. (p. 153)

4.5.1 Civil Disobedience

The fight against privatization elsewhere takes the form of civil disobedience in some of the well-known forms of land and food sovereignty. Civil disobedience as an act of resistance aims to secure additional rights usually through illicit actions that draw attention to the injustice of laws that forbid or criminalize things or states of existence that people are entitled to have (Bauer & Eckerstrom, 1987). In this context, the forms of power that are available to activists are nonviolent refusals to participate in modernity, development, or the capitalist privatization of space. Indeed, civil disobedience in the face of food system injustices is not new in the context of liberal democracies (Heynen, 2010).

The political will is informed by what Patel (2009b) calls a cosmopolitan federalism that recognizes a moral universal in the right to food. Actions or policies that undermine this right are considered immoral and unjust. Food sovereignty not only questions the morality of neoliberalism, according to Patel, but also calls for democratic political action to replace policies that privilege the powerful. There can be no doubt that the influence of transnational capital in the food system, particularly through the patenting and control of life forms, poses a potential threat to the sustainability of the food supply everywhere (Kloppenburg, 2010).

The saving of seeds in the neoliberal environments of India in a post-colonial development context is most certainly a political act, but Amy Trauger doubts whether it can be usefully framed as an act of civil disobedience. Trauger points out that Vandana Shiva makes an appeal to what Patel calls food sovereignty's moral universal right to food. While Navdanya opens up a space for criticism of India's policies on seeds, the association itself survives on the donations of charitable organizations, many of which are large corporations that receive enormous tax subsidies for giving away capital to organizations like Navdanya (Trauger, 2015). What Trauger is critical of is that Shiva strongly opposes corporations – including MNCs and TNCs – for taking away the rights of small farmers of India by patenting but her organization Navdanya accepts donations from large corporations for its survival. This in turn, perpetuates the existence and growth of corporations and serves a critical function for the global economy and defeats the purpose that Navdanya strives towards.

In the same vein, Brown (2018) points out that by accepting donations, Navdanya has 'become part of the machinery of hegemony: they help construct *consent* for the corporate food regime, rather than build radical alternatives to it' (p. 174). Brown shows that sustainable agriculture in India is paradoxical: the concept is ideologically opposed to the corporate food regime, but in practice, it is related to the institution that maintains the food regime (Brown, 2018). According to Brown, hegemony manifests itself in different ways, and what might appear to be radical alternatives may, in fact, involve cooperation with the ruling elite in a subtle way.

4.5.2 Human Right to Food

A noteworthy aspect of Navdanya is its emphasis on the human right to food. In their analysis of the case of Navdanya, Armaline and Glasberg show that Navdanya's partnership with small farmers, consumers, activists, and NGOs across the globe has helped the movement to respond to the corporate monopolization of food production and seed. Navdanya has participated in many campaigns by putting Shiva's ideas into practice. As a nuclear physicist, teacher, and activist, Shiva has documented and helped resist what she calls the corporate 'hijacking of the global food supply' (2000, 2001, 2008). Shiva describes how corporations such as Monsanto and Ricetec have been able – through trade agreements via the WTO (and its predecessor the General Agreement on Tariffs and Trade) and through

Structural Adjustment Programmes of the World Bank and the International Monetary Fund – to acquire land, develop monocultural (typically single crop) agricultural production, increase the use of genetically engineered crops and dangerous herbicides and pesticides, and patent (or 'hijack', as Shiva says) previously public crops developed over generations of farmers in agricultural communities primarily in the Global South. Armaline and Glasberg (2009) argue that 'as a result, large agribusiness and biotech firms have made great profits at the expense of agricultural communities forced into single-crop export production, if not forced off their land altogether' (p. 442).

Shiva's argument is that this process threatens the global biodiversity, the soil, and water revival processes as well as the culturally diverse relationships between humans and seeds that determine their co-existence (Shiva, 2000; 2008). Although companies like Monsanto claimed that they were fighting famine and increasing food supplies, Shiva points out that the corporatization of food production and the 'green revolution' promoting GMOs have taken away the right of a large number of people to produce and consume nutrient-rich, culturally appropriate, and ecologically sustainable food. It is, in Shiva's words, 'a recipe for starving people, not for feeding them' (2000, p. 13).

Companies like Monsanto, Bechtel, and Ricetec – which respectively have control over soy and pesticides, water, and rice – are only able to succeed in monopolizing global agriculture with the help of the governments of the countries in which they operate. By enforcing laws that protect these companies' patents, these governments in effect take into consideration the companies' need to maximize profits for their shareholders but turn a blind eye to human rights infringements (Armaline and Glasberg, 2009).

Through organized civil disobedience, lobbying, legal action, protests, and information-sharing, Navdanya and its many partners have been somewhat successful in resisting the more damaging effects of structural adjustment and corporate monopolization (Shiva, 2000, 2008). Armaline and Glasberg conclude that Navdanya's success has been due partly to its ability to target local, national, and transnational state structures as well as private TNCs.

Pritchard et al. notes that since the late 1990s, the concept of the right to food has gained political and ideological significance. It is a concept found in international treaties and humanitarian laws that asserts individuals' rights to live in dignity, free from hunger and malnutrition. Three core principles were specified by the United Nations Committee on Economic, Social and Cultural Rights in 1999: to respect, protect, and fulfill people's right to food. *Respecting* the right to food means not taking any measures that arbitrarily deprive people of their right to food. *Protecting* the right to food means enforcing laws and taking measures if third parties (including individuals and corporations) violate others' right to food. *Fulfilling* the right to food requires governments to be proactive by engaging in activities that aim to strengthen people's access to resources and to facilitate their ability to feed themselves. As Bill Pritchard et al. note:

> The fights-based perspectives on food are complemented by conceptualizing food security through the notions of freedoms and capabilities. The right to food is a hollow concept unless it is linked to the question of whether people are able to exercise, agitate and act to ensure this right is met. This brings into the frame the need for a *human-centred* focus. The issue of food security is understood not in terms of how much food is produced, but rather, whether and how those in need gain access to that food. (2014, p. 4)

They conclude that the food insecurity in India arises from 'a complex interplay of economic, social, political, environmental and cultural processes' (p. 17). Though in terms of economic development India might be better off, yet at the level of children and others suffering from malnutrition there is not much improvement. India does not follow the international trend of children underweight of age falling by roughly half the rate of GDP growth. India requires food policy solution which includes security of livelihood and an expanded notion of justice for the most defenseless population.

Human rights are mainly about upholding human dignity. To design sustainable food production, it is essential to have an understanding of the socio-cultural context. Human rights and sustainable development have to coexist and mutually reinforce each other. In their analysis of what they call 'human rights instruments', Armaline and Glasberg (2009) posit that one such instrument – the Universal Declaration of Human Rights – was built on the assumptions that (1) states have the ability or political will to fulfill their responsibilities detailed in human rights instruments to which they are party, (2) states are autonomous relative to each other and private interests, and (3) states actually represent and/or serve the interests of their general populations (p. 431). Their argument is that these human rights instruments present us with more challenges than benefits due to the global restructuring that took place as a result of the precedents set by efforts in the 1970s to resolve the crises in oil and other natural resources, the rise of large TNCs as the dominant corporate norm, the rise of transnational financial institutions or 'postnational finance capitalism' such as the International

Monetary Fund and the World Bank, and the rise of transnational trade organizations such as the World Trade Organization.

The point is that when there is a conflict between human rights instruments and the global restructuring features stated above, the state often takes the side of the TNCs with their motive of maximizing profits for their shareholders. Even though companies are legally obliged to honour the interests of their shareholders, in the context of food sovereignty, the state and the corporations deprive certain groups of people of their dignity by taking away the rights of the small farmers, women, and others who do not have much of a say. And depriving a group of people of their dignity is morally wrong, no matter what the circumstances are and no matter what the letter of the law is.

4.5.3 Protecting the Global South

At the international level, Navdanya has bonded with several movements for sustainability. It defends the Global South against the Global North by championing the small producers, particularly women, against large multinational corporations whose primary concern is making profit at any cost. By building its popular political and market base upon the Gandhian method of *satyagraha*, Navdanya has achieved international notoriety. Navdanya International has even opened an office in Florence, Italy. Vandana Shiva served as the Chair of the Florence office for ten years, collaborating with the regional government of Tuscany and the International Commission on the Future of Food and Agriculture. Navdanya has also partnered with non-Indian movements like Slow Food, an international movement promoting regional cuisine and local economic systems that enable the prosperity of small producers. Slow Food promotes the development of ethical consumption, biodiversity, and cultural diversity – ideas that match Navdanya's 'small is beautiful' philosophy perfectly. Gaia Foundation, based in the UK, is another international partner of Navdanya, founded in the mid-1980s by a group of ecological pioneers focusing on the global issues of environmentalism and biodiversity (Virmani & Lepineux, 2015). With its multifaceted strategy both within India and internationally, Navdanya has established itself on firm ground.

In addition to the Navdanya movement's success in stopping the abuse of power by large agribusinesses, it has also become a model for changing the fundamental social organization. Participants in grassroots organizations like Navdanya have engaged in decentralizing and democratizing their communities by organizing for their basic rights to sow and trade seed, collect clean water, and harvest food (Armaline & Glasberg, 2009). As Shiva (2008) explains:

> Navdanya builds community seed banks based on rescuing, conserving, reproducing, multiplying, and distributing native varieties or farmers' varieties – varieties evolved and bred over millennia. On the one hand, our seed saving defends seeds as a commons – resisting through our daily actions the degraded, immoral, uncivilized idea that seeds are the 'intellectual property' of corporations, and that saving them is a crime. On the other hand, Navdanya's seed banks are the basis of another food economy, one based on biodiversity and cultural diversity, on sustainability, and on the future. (p. 119)

There are some important points to remember here. Armaline and Glasberg show that due to its emphasis on human rights, Navdanya has succeeded in ensuring basic human dignities. Second, the Navdanya representatives illustrate that people can organize and govern themselves. Third, the struggle for human rights involves a larger project of confronting the authorities of the state and larger private corporations using the power of autonomous communities. In this confrontation, the local communities are capable of decentralizing by organizing and governing themselves and taking part in the decision-making process. Armaline and Glasberg point out that researchers and activists have also documented struggles against the privatization of water in places like Bolivia (Shiva, 2000; Dangl, 2007). Like the farmers of Navdanya and their allies, Bolivian activists resisted and overthrew state administrations that repeatedly partnered with corporations such as Bechtel to privatize all access to clean water. Similarly, these struggles over central resources quickly transformed themselves into broader struggles over wealth and power redistribution, culminating in a confrontation between the power of decentralized communities against the more highly centralized nation-state and corporate-owning class. 'A closer look at the human rights enterprise yields a pattern of struggle "from below" against powerful states and TNCs over the access to limited resources and political voice in decision making' (Armaline & Glasberg, 2009, p. 444).

Transnational parallels are abundant and staggering in their implications. As monocultures have replaced traditional polycultures, dependence on food and seed imported from international markets has increased, limiting food choices and food security. Food has become a transnational commodity and is the weapon deployed by multinational corporations to enforce what Shiva (2000) calls 'food totalitarianism' (p. 17) – the result of control over entire food chains by a few multinational corporations and the destruction of alternative, diverse, safe foods produced ecologically.

4.6 Shiva's View of Earth Democracy

In order to showcase the profound philosophical – that is, ethical and political – dimensions of Shiva's work, I explore here in some depth Hrynkow's analysis of Shiva's three recent books, *Earth Democracy: Justice, Sustainability and Peace* (2016a); *Soil, Not Oil: Climate Change* (2016b); and *Who Really Feeds the World* (2016c). These works capture the essence of Shiva's philosophy of Earth Democracy, the principles of which underpin the practices of Navdanya.

Earth Democracy is premised on the basic idea that all things are interconnected. This holistic orientation advances an inherently relational worldview supported by ecological science and Shiva's knowledge of quantum theory (her doctorate was on quantum theory from the Western University in London, Ontario). Shiva extends the interconnectedness of the sciences of ecology and quantum mechanics to Mother Earth or the planet as a whole. This relational worldview, according to Shiva, is also present in Pope Francis' encyclical, *Laudato si'* (2015), an ecclesiastical document that highlights the importance of caring for our common home. Shiva (2016a) recognizes a convergence here with the Indian philosophy of *Vasudhaiv Kuttumbakin*, which translates as 'the world is one family'. In this worldview, all beings have rights (Hrynkow, 2018). Some have emphasized that the attribution of rights to all beings carries eco-spiritual connotations (see, for example, Berry, 2006).

Hrynkow mentions Shiva's argument that these intrinsic rights have been taken away by the violence of colonialism and corporate globalization. In particular, she frequently returns to the violence meted out by the Green Revolution, manifested by chemically intensive industrial agriculture and the privatization and bottling of community water supplies by multinational corporations such as Coca-Cola (Shiva, 2016a). Hyrnkow says that for Shiva, such violence is only brought into the realm of the possible by a de-sacralization of nature and the removal of the earth's status as a mother (Shiva, 2016c; Hrynkow, 2018).

Hyrnkow also points out that in tracing how the Green Revolution's violence to nature took place, Shiva relies on the views of two ecofeminists, Carolyn Merchant and Val Plumwood. Shiva is critical of those modern Western scientists and philosophers who were shortsighted about the interconnectedness of man and nature and who viewed nature as subordinate – in particular, Rene Descartes with his dualistic worldview, Francis Bacon with his notion that feminine nature had to be subordinate, Isaac Newton with his proposal of a mechanistic universe, and Charles Darwin with his view of nature being conflict-ridden and not cooperative (Shiva, 2016c; Hrynkow, 2018).

By contrast, Earth Democracy seeks to recover and revitalize a more holistic worldview that understands how the social, political, and ecological dimensions are all intertwined within webs of intractable relationships. Elegantly put, Shiva recommends that political ethics act as an enabling mechanism permitting 'us to reclaim our common humanity and our unity with all life' (2016a, p. 8). This goal is supported in Shiva's ten principles of Earth Democracy. In harmony with the Earth Charter Commission (2000), the first principle assigns intrinsic worth to all species, peoples, and cultures. The second principle – recalling the assertion by Leonardo Boff (2011) that the proper governance regime for the earth community is 'a socio-cosmic democracy' – emphasizes the need for a democracy to be accountable to all living beings. It is our duty to protect the ecological processes and the rights of all people and other living beings. Invoking the intimate connection between culture and nature (see, for example, Lake, 2010), the third principle of Earth Democracy articulates a duty to defend nature and culture, as they are the basis of sustainability and peace. The fourth principle asserts that all beings have the right to sustenance.

The natural right to life itself cannot be eroded. In accordance with Sallie McFague's articulation (2001) of imperatives to rethink values and economy in a time of planetary peril, the fifth and sixth principles uphold the importance of local, democratic, and living economies that provide for the basic needs of all and only trade across long distances for goods that cannot be produced locally. The localization of economies is a social and ecological imperative. The seventh principle affirms the value of living democracies that are rooted in participation and subsidiarity (compare Eckersley [2004] on the qualities of the green state). Local communities must have a role to play in decision-making. In line with many indigenous socio-politically oriented thinkers (for example, Winona LaDuke), the eighth and ninth principles declare that the coupling of nature with vibrant and diverse living cultures is essential for Earth Democracy. Living cultures allow for cultural diversity and recognize the multiplicity of identities. Finally, the tenth principle ties Earth Democracy to an alternative world system that 'globalizes compassion, justice, and sustainability' (Shiva, 2016a, pp. 9-11). Ultimately, Earth Democracy unites and does not divide (Hrynkow, 2018).

Shiva compares this potential of Earth Democracy with the limited achievements of even the best climate change mitigation proposals generated through the Conference of the Parties processes within the United Nations Framework Convention on Climate Change, such as the greatly praised plans put forward at the Copenhagen meetings in 2009. In Shiva's view, the Copenhagen plans only extend the lifespan of an exploitative economic model. In contrast, she argues that Earth Democracy, based as it is on principles of justice and sustainability, 'can help us build another future for the human species – a future in which we recognize we are members of the earth family, that protecting the earth and her living processes is part of our species' identity and meaning' (2016b, pp. xvi-xvii). Agriculture is a key link in this eco-political programming because, according to Shiva, 'Earth Democracy grows in the fertile soil shaped by the earth, the human imagination and human action' (2016b, p. 7). Hrynkow (2018) submits that 'It is this understanding that in a relevant way spiritualizes her politically charged project, with all the tensions and advantages inherent therein, and shapes her treatment of ecological agriculture' (p. 208).

Shiva's vision of Earth Democracy is grounded in her articulation of agroecology. This approach to sustainable agriculture is explicitly centred on the local. Agroecology seeks to transform trends in global food production in order to support small farmers along with cultural, crop, and biological diversity (Shiva, 2016b). To understand the transformation that Shiva is advocating in this regard, it is necessary to understand her criticisms of high-intensity, chemically based industrial agriculture. She argues that violence and exploitation are chemically based, and industrial agriculture comes from the foundation of violence and exploitation. She points out that chemical herbicides and pesticides, while directed against creatures that ravage crops, have ended up attacking the ecosystem. This kind of assault takes place only when the inherent link within the earth community is denied due to the dualistic and mechanistic worldview supported by Descartes, Newton, Bacon, and Darwin (Shiva, 2016c). Hence her plea for a transformation to agroecology based on the foundation of sustainable agriculture (Hrynkow, 2018).

Shiva's idea of Earth Democracy has a direct correlation with Navdanya. On Navdanya's website, there is a section on Earth Democracy that mentions the ideals of seed sovereignty (*bij swaraj*), food sovereignty (*anna swaraj*), water sovereignty (*jal swaraj*), and land sovereignty (*bhu swaraj*). Shiva argues that a paradigm shift has to be made to view these four ideals as parts of a whole.

4.7 Genetically Modified Crops and the Future

Climate change and food crises are two major issues that policymakers and scientists are concerned about globally, and many are looking to genetically modified (GM) crops for solutions. Scientists in India believe that the future of crops looks favourable, provided some policy changes are made. Over the last two decades, research about GM crops as well as their evaluation and monitoring have advanced significantly. Nonetheless, there is yet no consensus within the Indian government and among ministries and departments of the regulatory system on how to overcome the food and climate challenges (Shukla et al., 2018).

Many scientists argue that agricultural biotechnology – and GM technology – must be adopted by India, not only for the sake of the country but also to address the issue of the global water crisis and to combat climate change. India uses more than 80 percent of its surface water for agricultural use; industry and household use account for only 20 percent of water resources. According to Bajaj (2017), 'India is also the largest user of groundwater in the world, with over 60 percent of irrigated agriculture and 85 per cent of drinking water supplies dependent on aquifers'. He argues that policymakers need to reevaluate the long-term relevance of agricultural biotechnology. Though GM technology might not be the only way to solve all the problems related to the water crisis and climate change, Bajaj makes the case that this should not be the reason to exclude GM technology. All safe technologies should be adopted, which would allow the farmers themselves to make the decisions that are best for them (Bajaj, 2017).

In India, it is projected that the availability of water for agriculture will decline considerably, as the per capita water availability has already declined from 5,000 cubic metres per annum in 1950 to around 2,000 now and is projected to be 1,500 by 2025 (Bajaj, 2017).

It is important to note here that the Indian government is slowly moving towards GM crops even in the face of opposition from Gene Campaign (an NGO advocating biodiversity), Swadeshi Jagran Manch (SJM) – an affiliate of Rashtriya Swayamsevak Sangh (RSS) (anti-GM crop environmentalists) – and other environmental movements in India. The central government's research institution, the Indian Council of Agricultural Research (ICAR), is joining the Vasantdada Sugar Institute (VSI) led by former agriculture minister Sharad Pawar to develop drought-resistant sugarcane that will need much less water than non-GM sugarcane, which uses a considerable amount of water. In Maharashtra, the sugar cane crop is a major cause of the water crisis (Mohan, 2016). The states of Karnataka and Maharashtra are both suffering from a shortage of drinking water and financial losses from cash crops. In Maharashtra, 10 percent of the cultivated area is used for sugarcane, but it consumes three-fourths of all water used in agriculture (Gokhale, 2017).

ICAR scientists believe that developing GM sugarcane that need less water will be a long-term project. One scientist said, 'We know how difficult it is in India to go for commercial release of transgenic crop' (Mohan, 2016). Nonetheless, there has been a discussion between the chief minister of Maharashtra, Devendra Fadnaviss, and Prime Minister Modi about bringing all the sugarcane-growing area under drip irrigation within three years. It is a long process to finally get approval from the central regulator, the Genetic Engineering Approval Committee (GEAC) of the environment ministry.

The current Modi government is supportive of GM crops, but some state governments are not willing to embrace this technology (Kumar, 2015). In 2017, the GEAC recommended the commercial cultivation of GM mustard (Brassica juncea) researched at the University of Delhi, but the final decision is still pending (*Indian Express*, 2017). Separately, India's Supreme Court heard a case brought by Aruna Rodrigues, an anti-GM campaigner who has questioned GEAC's transparency, as the scientists were accused of deceiving the public about the benefits of transgenic mustard (Kumar, 2017). The patent for this mustard will remain with the Indian government, unlike the Bt cotton patent, which is owned by private industry.

Scientists and the central government have been working in a dedicated way to get approval of some GM crops. One of the oldest examples is Mahyco (Maharashtra Hybrid Seed Company), which developed Bt brinjal, or eggplant, in 2000. Bt brinjal was approved by GEAC in 2010, but recent concern about its safety and efficacy prompted Environment Minister Jairam Ramesh to put it on indefinite moratorium. The states and the parliament are critical about the regulatory system. When Bt brinjal was approved for commercial use in Bangladesh in 2013 and the crops lost their resistance to pests a few years later, officials nonetheless maintained that it is a success in Bangladesh (Sushma, 2019). On biosafety issues, scientists take stands on both sides, some in favour of releasing Bt brinjal and others pointing out problems with the characterization of Bt brinjal and its environmental impact assessment. As Aniket Aga (2019) recounts, 'The ecologist, Madhav Gadgil, warned of contamination of India's diverse brinjal varieties. Biodiversity is critical for nutrition and sustainability, and the government's own taskforce on biotechnology (2004) had recommended that no GM crop be allowed in biodiversity-rich areas'. As Bt brinjal still does not fit into the framework of agricultural development and farmers' well-being as stated by the parliamentary panels and other expert committees, its cultivation has been suspended since 2010.

Elsewhere, the International Rice Research Institute (IRRI) is developing rice varieties that can tolerate the intense effects of climate change like drought, flooding, heat, cold, and soil problems including high salt and iron toxicity. Severe drought in some states in India can cause a 40 percent loss in rice yield, equivalent to a loss of $800 million. Several varieties of drought-tolerant rice have been released and are being planted by farmers, including Sahbhagi Dhan in India, Sahod Dhan in the Philippines, and the Sookha Dhan varieties in Nepal. IRRI scientists have identified some key regions of the rice genome, called quantitative trait loci (QTLs), that produce drought tolerance and high yields of rice. The scientists are working to introduce drought tolerance to the popular high-yielding rice varieties including IR64, Swarna, and Vandna (see IRRI website). India has also recently launched an improved chickpea through genome-assisted breeding, developed in 2014 by the Agricultural Research Institute of India and Raichur (Karnataka) University of Agricultural Sciences in collaboration with the international organization International Crops Research Institute for the Semi-Arid Tropics (ICRISAT). ICRISAT is a non-profit, non-political organization that conducts agricultural research for development in the drylands of Asia and Sub-Saharan Africa. This variety of chickpea, called Pusa 10216, has exhibited high tolerance to drought. Chickpea is a high-protein vegetarian food that is consumed by many Indians. India is the largest producer of chickpeas in the world and the largest importer (see Drishti website). It is hoped that this research will help the significant decline in productivity in the production of chickpeas in India.

Those who remain opposed to GM technology argue that such technology, when controlled by powerful multinational corporations, takes away the rights of small farmers in India, reduces the biodiversity of the ecosystem, and raises possible health issues like allergies and resistance to antibiotics among consumers. Pro-GM scientists argue that all the problems mentioned by anti-GM environmentalists can also occur with non-GM crops. But if climate change is causing water shortage, and given that certain GM crops consume huge amounts of water, which one is the better option? Of the conflicting concerns related to how the country copes with the water crisis – loss of biodiversity and the lack of socio-economic benefits for small farmers – which one should take precedence? Clearly, this is a moral dilemma that is very complex.

The opponents of GM crops argue that it is better to develop hybrid varieties that can tolerate the drought. The Modi government is pushing for the approval of GM mustard, but no data has been provided yet on whether GM varieties are better than hybrid varieties. Opponents say the Modi government is more interested in obtaining the proprietary rights of the GM technology. It will be interesting to see how GM crop technology develops in the future in India and globally.

4.8 Conclusion

This chapter first set up the context in which the debate over genetically modified organisms (GMOs) has taken place. It then provided the background of the Green Revolution, which started with the idea of developing the most efficient system of food productivity and ended up triggering resistance by anti-GM activists who argue that the Green Revolution led to the monoculture of crops, had a negative impact on biodiversity, and eliminated the choices that were available for local small farmers. The Green Revolution has encouraged multinational and transnational corporations to take over the agriculture of developing countries like India. The different protest movements such as the Gene Campaign, the KRSS, and Navdanya, have been at the forefront of the anti-GM movements in India by challenging the economic development model based on efficiency alone without considering sustainable agriculture, food sovereignty, and biodiversity in addressing the issue of climate change.

This chapter has focused on Navdanya, whereby the movement's three different aspects were analyzed – human rights, nonviolence, and ecofeminism – within the broader umbrella of sustainable agriculture, biodiversity, and culture. In the area of human rights, Navdanya has fought for the right to food as part of the struggle for food sovereignty. It has done so using a bottom-up approach of giving priority to local farmers and communities and by protesting against the top-down management approach of multinationals and the state, which, when working together legally on the basis of the sole motive to increase profits, inevitably end up exploiting local famers and communities.

In terms of civil disobedience, Navdanya has adopted the Gandhian method of *satyagraha* to make a political statement about social justice. Shiva's critics have argued that although Shiva has introduced a different political discourse, she may nevertheless have fallen into the same trap by accepting donations from large corporations. Thus, although Navdanya is ideologically opposed to the corporate food regime, in practice, it has joined hands with it.

In this chapter I have also explored the connection between ecofeminism and environmental justice as well as a specific version of ecofeminism, namely, materialist ecofeminism, highlighting Shiva's vision of Earth Democracy and its ten principles. Shiva and in turn Navdanya have made significant contributions to the discourse on sustainable development about GM crops. In the concluding part of this chapter, the other side of the issue of GM crops was presented by showing how agricultural biotechnology is moving forward slowly in India; it remains to be seen what shape it will take.

In the next chapter, I examine the moral and theoretical framework of the three movements discussed in this book, showing how the power of nonviolence, the six core values of development ethics, the ethics of care, and southern materialist ecofeminism have been applied to these environmental movements. In essence, Chipko, NBA and Navdanya all have moral foundations.

Works Cited

Aga, Aniket (2019). 'Serious concerns over Bt brinjal', comment in *The Hindu*, 18 June. Available at https://www.thehindu.com/opinion/op-ed/serious-concerns-over-bt-brinjal/article28022577.ece (accessed 12 August 2019).

Arendt, Hannah (1967). *The Origins of Totalitarianism*. Allen & Unwin: London.

Armaline, William T., and Davita Silfen Glasberg (2009). 'What Will States Really Do for Us? The Human Rights Enterprise and Pressure from Below'. *Societies Without Borders*, 4, 430-451.

Artner, Annamaria (2004). 'Anti-globalization Movements: The Developments in Asia'. *Contemporary Politics*, 10, 3-4, 243-255.

Assadi, M. (2004). 'Social Movements in Karnataka'. In J. Desrochers and P.V. Veliyannoor (eds.), *Poverty, Marginalization and Empowerment in Karnataka*. Bangalore: KRCR-JPC Publications, 197-215.

Bajaj, Shivesh (2017). Opinion: *GM crops, The Answer to Climate Change* https://www.thehindubusinessline.com/opinion/gm-crops-the-answer-to-climate-change/article9711720.ece (accessed 24 May 2017).

Bauer, S.M., and P.J. Eckerstrom (1987). 'The State Made Me Do It: The Applicability of the Necessity Defense to Civil Disobedience'. *Stanford Law Review*, 1173-2000.

Berry, T. (2006). *Evening Thoughts: Reflecting on Earth as Sacred Community*. San Francisco: Sierra Club Books.

Boff, Leonardo (2011). *Virtues for Another Possible World*, translated by A. Guilherme. Eugene, OR: Cascade Books.

Brown, Trent (2018). *Farmers, subalterns, and activists: Social politics of sustainable agriculture in India*. Cambridge: Cambridge University Press.

Croeser, Sky (2015). *Global Justice and the Politics of Information: The Struggle over Knowledge*. London: Routledge.

Dangl, Benjamin (2007). *The Price of Fire: Resource Wars and Social Movements in Bolivia*. Oakland, CA: AK Press.

Drishti (2019). 'Climate-Resistant Chick Pea Varieties', April 30. https://www.drishtiias.com/printpdf/climate-resistant-chickpea-varieties (accessed 15 June 2020).

Earth Charter Commission (2000). The Earth Charter available at: http://earth-charter.org/discover/the-earth-charter/ (accessed 20 January 2019).

Eckersley, Robyn (2004). *The Green State: Rethinking Democracy and Sovereignty*. Cambridge, MA: The MIT Press.

Etmanski, Catherine (2018). 'Seeds and Stories of Transformation from the Individual to the Collective'. *Journal of Transformative Education*, 16, 2, 151-167.

Francis (2015). *Laudato si'*. Vatican City: Vatican Press.

Gokhale, Nihar (2017). 'Can GM Sugarcane really make a difference in drought-hit areas'? http://www.catchnews.com/environment-news/can-gm-sugarcane-really-make-a-difference-in-drought-hit-areas-1462379669.html (accessed 2 July 2020).

Government of India (2000). *National Agriculture Policy – 2000*. New Delhi: Ministry of Agriculture. https://www.indiawaterportal.org/articles/ministry-agriculture-announces-national-agricultural-policy-2000

— (2002). *Tenth Five Year Plan 2002-2007*, Vol. II. New Delhi: Planning Commission.

Gruere, Guillaume P., Purvi Mehta-Bhatt, and Debdatta Sengupta (2008). 'Bt Cotton and Farmer Suicides in India: Reviewing the Evidence'. *IFPRI Discussion Paper n 808*, Washington, DC: International Food Policy Research Institute.

Gupta, Anil, and Vikas Chandak (2005). 'Agricultural Biotechnology in India: Ethics, Business and Politics'. *International Journal of Biotechnology*, 7, 2, 212-227.

Harms, Arne, and Oliver Powalla (2014). 'India: The Long March to a Climate Movement'. In Dietz Matthias and Heiko Garrelts (eds.), *Routledge Handbook of Climate Change Movement*. London: Routledge, 179-193.

Herring, Ronald J. (2008). 'Whose Numbers Count? Probing Discrepant Evidence on Transgenic Cotton in the Warangal District of India'. *International Journal of Multiple Research Approaches*, 2, 2, 45-59.

— (2015). 'State Science, Risk and Agricultural Biotechnology: Bt Cotton to Bt Brinjal in India'. *Journal of Peasant Studies*, 42, 1, 159-186.

Heynen, Nik (2010). 'Cooking up Non-Violent Civil-Disobedient Direct Action for the Hungry: "Food Not Bombs" and the Resurgence of Radical Democracy in the U.S.' *Urban Studies*, 47, 6, 1225-1240.

Hrynkow, Christopher (2018). 'Situating Earth Democracy: Vandana Shiva on Agroecology, Contemporary Politics and Resilience'. *Political Studies Review*, 16, 3, 205-216.

Indian Express (2017). 'Genetically-modified mustard gets GEAC nod for cultivation'. https://indianexpress.com/article/India/gm-genetically-modified-mustard-gets-geac-genetic-engineering-appraisal-committee-nod-for-cultivation-4651857/ (accessed 29 June 2020).

IRRI. https://www.irri.org/climate-change-ready-rice (accessed 1 July 2020).

Kloppenburg, Jack (2010). 'Impeding Dispossession, Enabling Repossession: Biological Open Source and the Recovery of Seed Sovereignty'. *Journal of Agrarian Change*, 10, 3, 367-388.

Krishnakumar, Asha (2005). 'Seeds of Controversy'. *Frontline*, 22, 12, 4-17.

Kumar, Sanjay (2015). 'India eases stance on GM crop trials'. *Nature*, 521, 138-139.

— (2017). 'India's GM mustard on hold'. *Nature*, 541, 267-268. https://www.nature.com/articles/541267a.pdf?origin=ppub (accessed 10 May 2019).

Kumari, Madhulika, and Sambit Mallick (2015). 'Bacillus Thuringiensis Cotton in India: Debates Surrounding Agricultural Biotechnology'. In Satyajit Majumdar, Samapati Guha, and Nadiya Marakkath (eds.), *Technology and Innovation for Social Change*. New Delhi: Springer, 161-255.

LaDuke, Winona (1999). *All Our Relations: Native Struggles for Land and Life*. Cambridge, MA: South End Press.

Lake, Osprey Orielle (2010). *Uprisings for the Earth: Reconnecting Culture with Nature*. Ashland, OR: White Cloud Press.

Lele, Sharachandra (2012). 'Climate Change and the Indian Environmental Movement'. In N. Dubash (ed.), *Handbook of Climate Change and India: Development, Politics and Governance*. London & New York: Earthscan.

Mahadevan, Renuka (2003). 'Productivity Growth in Indian Agriculture: The Role of Globalization and Economic Reform'. *Asia-Pacific Development Journal*, 10, 1, 57-71.

McFague, Sallie (2001). *Life Abundant: Rethinking Theology and Economy for a Planet in Peril*. Minneapolis, MN: Augsburg Fortress.

Merchant, Carolyn (1980). *Death of Nature: Women, Ecology and the Scientific Revolution*. New York: Harper Collins.

Mishra, Srijit (2014). 'Farmers' Suicides in India, 1995-2012: Measurement and Interpretation'. *Asia Research Centre Working Paper*. London: London School of Economics & Political Science.

Mohan, Vishwa (2016) 'ICAR looks at GM sugarcane option to ease water woes', 8 May, *Times of India*, https://timesofindia.indiatimes.com/home/science/ICAR-looks-at-GM-sugarcane-option-to-ease-water-woes/articleshow/52179120.cms (accessed 2 July 2020).

Monshipouri, Mahmood, Claude E. Welch Jr., and Evan T. Kennedy (2003). 'Multinational Corporations and the Ethics of Global Responsibility: Problems and Possibilities'. *Human Rights Quarterly*, 25, 965-989.

Nadal, Alejandro (2007). 'Monsanto, Cereal Killer GM and Agrarian Suicides in India', 6 January, *La Jornada*.

Nanjundaswamy, M.D. (1998). 'Cremating Monsanto: Genetically Modified Fields on Fire'. In Notes from Nowhere (ed.), *We are Everywhere: The Irresistible Rise of Global Anticapitalism*. London and New York: Verso, 152-154.

Navdanya (2007). *The Seed and the Spinning Wheel: Two Decades of Swaraj & Satyagraha for Seed Freedom*. New Delhi: India, Navdanya/RFSTE.

— (2007). *Cargill: The Corporate Hijack of India's Food and Agriculture*. New Delhi: Navdanya/RFSTE.

— (2008a). 'About us,' Navdanya, available online at https://www.navdanyainternational.it/en/about-us-navdanya-international (accessed 28 October 2018).

Newell, Peter (2007). 'Biotech Firms, Biotech Policies Negotiating GMOs in India'. *Journal of Environmental Development*, 16, 2, 183-206.

Ninan, K.N., and H. Chandrashekar (1993). 'Green Revolution, Dry Land Agriculture and Sustainability: Insights from India'. *Economic & Political Weekly*, 12, A2-A7.

Nyéléni Declaration (2007). 'Proceedings of the Forum for Food Sovereignty Held in Selengue, Mali, February 23-27'. Also available at http://usfoodsovereigntyalliance.org/what-is-food-sovereignty/.

Patel, Raj (2009a). 'Food Sovereignty: A Brief Introduction'. Available at http://rajpatel.org/2009/11/02/food-sovereignty-a-brief-introduction/.

— (2009b). 'What Does Food Sovereignty Look Like?'. *The Journal of Peasant Studies*, 36, 3, 663-706.

Peschard, Karine (2014). 'Farmers' Rights and Food Sovereignty: Critical Insights from India'. *The Journal of Peasant Studies*, 41, 6, 1085-1108.

Philpott, Tom (2015). 'No, GMOs Didn't Create India's Farmer Suicide Problem, But...'. *Mother Jones*, September 30. www.motherjones.com/tom-philpott/2015/09/no-gmos-didnt-create-indias-farmer-suicide-problem (accessed 16 March 2019).

Plumwood, Val (1993). *Feminism and Mastery of Nature*. London: Routledge.

Pritchard, Bill, et al. (2014). *Feeding India: Livelihoods, Entitlements and Capabilities*. Routledge (Earthscan).

Rawat, Raghubir Singh (2010). 'Participatory Research on Impacts of Climate Change on People's Livelihood in Garhwal Himalaya'. In Vandana Shiva and Vinod Kumar Bhatt (eds.), *Climate Change at the Third Pole*. New Delhi: Navdanya.

Sahai, Suman, Bhaskar Mahanta, and Waquar Ahmed (2005). *Farmers' Perception of Agrobiodiversity: A Study in North India*. New Delhi: Gene Campaign.

Scoones, Ian (2003) 'Regulatory Maneuvers: The Bt Cotton Controversy in India'. Sussex: Institute of Development Studies.

— (2005). 'Contentious Politics, Contentious Knowledges: Mobilizing Against GM Crops in India, South Africa and Brazil'. Sussex: Institute of Development Studies.

— (2008). 'Mobilizing Against GM Crops in India, South Africa and Brazil'. *Journal of Agrarian Change*, (2 & 3), 315-344.

Shiva, Vandana (2000). *Stolen Harvest: The Hijacking of the Global Food Supply*. Boston, MA: South End Press.

— (2001). *Water Wars*. Boston, MA: South End Press.

— (2008). *Soil, not Oil*. Boston, MA: South End Press.

— (2010). *Violence of the green revolution*. New Delhi: India, Natraj.

— (2014). *The Vandana Shiva Reader*. Foreword by Wendell Berry. Lexington, KY: University Press of Kentucky.

— (2016a). *Earth Democracy: Justice, Sustainability and Peace*. London: Zed Books.

— (2016b). *Soil, not Oil: Climate Change, Peak Oil*. London: Zed Books.

— (2016c). *Who Really Feeds the World?* London: Zed Books.

—, and Kunwar Jalees (2006). *Seeds of Suicide: The Ecological and Human Costs of Globalization and Agriculture*. New Delhi: India, Navdanya.

—, and Vinod Kumar Bhatt (2009). 'Himalayan Ecosystems and Climate Change: An Overview'. In Vandana Shiva and Vinod Kumar Bhatt (eds.), *Climate Change at the Third Pole*. New Delhi: Navdanya.

Shukla, Manish, et al. (2018). 'Status of research, regulations and challenges for genetically modified crops in India'. *GM Crops & Food*, 9, 173-188.

Stone, Glenn Davis (2004). 'Biotechnology and the Political Ecology of Information in India'. *Human Organization*, 63, 2, 127-140.

Sushma, Meenakshi (2019). 'Damming allegations: India's fields still have Bt brinjal'. *Down to Earth*, news, agriculture, 1-4.

Thomas, Gigesh, and Johan De Tavernier (2017). 'Farmer-suicide in India: Debating the role of biotechnology'. *Life Sciences, Society and Policy*, 13, 8, 1-21. Open Access. DOI 10.1186/s40504-017-0052-z.

Trauger, Amy (2015). 'Seed Sovereignty as Civil Disobedience in Northern India?' In Amy Trauger (ed.), *Food Sovereignty in International Context: Discourse, Politics and Practice of Place*. Routledge, 106-124.

Tripathi, Amarnath, and A.R. Prasad (2009). 'Agricultural development in India since independence: A study on progress, performance, and determinants'. *Journal of Emerging Knowledge on Emerging Markets*, 1, 1, 63-92.

Virmani, Arundhati, and Francois Lepineux (2015). 'Spiritual-Based Entrepreneurship for an Alternative Food Culture: The Transformational Power of Navdanya'. In Laszio Zsolnai (ed.), *The Spiritual Dimension of Business Ethics and Sustainability*. Cham: Springer International Publishing, 125-142.

Vyas, Vijay S. (2003). *India's Agrarian Structure, Economic Policies and Sustainable Development*. New Delhi: Academic Foundation Publishers.

5 Moral Implications of Environmental Movements

Abstract

This chapter delves into the importance of the well-being of all human beings, including the voices of the marginalized people in the community who are directly and negatively affected by deforestation, dam-building, and GMOs in India. It has been argued that Gandhi was not an environmentalist, although he believed in the oneness of nature. The chapter also considers the six core values that development ethicists have proposed, highlighting the moral aspect and the gender bias of the resettlement and rehabilitation policies aimed at the dispossessed people of the Narmada case. The chapter concludes with an analysis of the Chipko movement's connection to ecofeminism and the ethics of care.

Keywords: development ethics, globalization, southern materialist ecofeminism, human well-being, resettlement and rehabilitation, gendered environmental justice

5.1 The Mesmerizing Power of Nonviolence

Mahatma Gandhi's life and work have had a considerable influence on environmental movements in India. In one of the first printed accounts of the Chipko movement, 'a breathless journalist announced that Gandhi's ghost had saved the Himalayan trees' (Guha, 1995, p. 47). Since that time, Gandhi has been the typically acknowledged patron saint of environmental activists, many of whom rely greatly on Gandhian techniques of nonviolent protest and draw extensively on Gandhi's polemic against heavy industrialization. Some of the Chipko movement's better-known figures, including Chandi Prasad Bhatt, Sunderlal Bahuguna, and Baba Amte, have repeatedly underlined their debt to Gandhi (Guha, 1995). This chapter presents the moral and philosophical underpinnings of environmental movements – from Gandhi's

Mallick, Krishna, *Environmental Movements of India: Chipko, Narmada Bachao Andolan, Navdanya*. Taylor & Francis Group, 2021.
DOI: 10.5117/9789462984431_CH05

satyagraha to globalism, from the theory of development ethics as applied to the Narmada case to the burgeoning ecofeminist movements that remind us of our interdependence with natural resources as well as with each other.

Nonviolence, as envisioned by Gandhi, has the power to solve many of our problems, including ecological crises. Indeed, Gandhian nonviolence has been accepted by many different environmental movements as a vital principle. These movements have laid claim to Gandhian values of ecological prudence and frugality as well as decentralized democracy through village *swaraj,* or self-rule. But it is a type of nonviolence that has the mesmerizing effect of attracting the minds of human beings and of appealing to the heart. In Gandhi's own words:

> Suffering is the law of human beings; war is the law of the jungle for converting the opponent and opening his ears, which are otherwise shut, to the voice of reason. Nobody has probably drawn up more petitions or espoused more forlorn causes than I, and I have come to this fundamental conclusion that if you want something really important to be done, you must not merely satisfy the reason; you must move the heart also. The appeal of reason is more to the head, but the penetration of the heart comes from suffering. (*Young India*, 1931)

Gandhi's view of truth is connected to his religious beliefs and practices. As Terchek (2011) explains, 'it is a cosmological view of our relationship of one another and a pervasive sense of duty we owe to one another' (p. 118). His faith in nonviolence and commitment to vegetarianism made him a votary of the conservation of diversity, including all forms of life, societies, cultures, religions, and traditions. He made manifest the internal relation between self-realization, nonviolence, and what has sometimes been called *biospherical egalitarianism.*[3] Gandhi explained his concept of nonviolence as follows in the weekly journal *Harijan:*

(1) Nonviolence is the law of the human race, and is infinitely greater than and superior to brute force.
(2) In the last resort, it does not avail to those who do not possess a living faith in the God of Love.

3 By 'deep ecology', Arne Naess means that everything within the ecosystem has an inherent value irrespective of the needs for humans. Biospherical egalitarianism, as a principle of deep ecology, according to Naess, refers to the normative aspect of justice between human and non-human species. Terchek refers to the normative aspect.

(3) Nonviolence affords the fullest protection to one's self-respect and sense of honour.
(4) Individuals and nations who practice nonviolence must be prepared to sacrifice everything for the welfare of the whole world.
(5) Nonviolence is a power which can be wielded equally by all – children, young men and women, or grown up people, provided they have a living faith in the God of love, and therefore have equal love for all mankind. When nonviolence is accepted as the law of life, it must pervade the whole being and not merely be applied to isolated acts.
(6) It is a profound error to suppose that whilst the law is good enough for individuals, it is not for masses of mankind. (5 September, 1926, pp. 236-237)

5.1.1 An Ecological Warrior

Scholars such as Godrej (2012) argue that in order to address the issue of environmentalism in relation to Gandhi's nonviolence, it is important to understand the two different facets of Gandhi's nonviolence: the ascetic or ethical (*Brahmin*) aspect and the warrior-like (*Kshatriya*) aspect. Godrej notes that many environmentalists are truly inspired by the ascetic elements of Gandhi's project, a dimension that involves living simply and minimizing material consumption by following the ethics of self-denial and self-scrutiny. But she points out that in doing so, they do not take into consideration the warrior-like political nonviolence that Gandhi used, side-by-side with his asceticism. Gandhi interweaved the ethical facet with the strategic political facet. It was confrontational, assertive, and at times disruptive, while not using any form of violence. According to Godrej:

> Gandhi can thus be read as an advocate of a certain form of 'ecological' citizenship, focused on the fulfillment of obligations with respect to the political community, a fulfillment which requires both the scrutiny of one's bodily consumptive behaviours, as well as the necessary and literal placement of one's body on the frontlines of aggressive political contestation. (2012, p. 438)

Many who perhaps see themselves as acting within Gandhi's legacy may miss the point of this political assertiveness, expressed through the bodily dimension of political action. Contemporary environmental movements tend to overemphasize the self-abnegating, self-denying, and self-scrutinizing ascetic components of Gandhi's thought to the neglect of the confrontational and warrior-like ones. In so doing, they also frequently overemphasize the ethical dimension, missing the discursive political dimension with which this Gandhian ethics is tightly interwoven (Godrej, 2012).

The environmentalists of today do not merely claim that they are following the example of Gandhi, they go on to argue that the Mahatma himself foresaw the ecological crisis of modern industrial society. Yet while the question of whether Gandhi was indeed an early environmentalist is usually answered in the affirmative by his admirers, it is rarely backed up with supporting evidence. Guha (1995) asserts: 'it is taken for granted that Gandhi anticipated our environmental concerns, but without demonstrating precisely where and in what ways he did so' (p. 48). Some scholars assume that since Gandhi's life and writings have served as inspiration for many environmentalists, he must therefore have naturally been concerned with environmental preservation and sustainability. In fact, Godrej (2012) points out that Gandhi was remarkably silent on the question of the inherent value of nature or of human beings' relationship to their external environment: 'It is important, therefore, to identify Gandhi's position (if any) on the environment, in contrast to the assumptions often attributed to him, but seldom supported by textual evidence' (p. 438). It can be said that although Gandhi's writings have strong environmental implications, there is simply not enough textual evidence to prove that Gandhi was indeed an environmentalist (Godrej, 2012).

If his writings are tendered as evidence of Gandhi being an environmentalist, it is almost always his book *Hind Swaraj* (published in 1909) that is referred to. This work gives us an 'alternate perspective' on development (Kumar, 1992, p. 294) while explaining how 'the current mode of development is exploitative of man by man and of nature by man' (Guha, 1995, 48). Guha points out that although *Hind Swaraj* contains an eloquent denunciation of modern Western culture, it does not have anything to say about humans' relationship with nature and has even less to offer as an alternative perspective on development (Guha, 1995; Guha & Martinez-Alier, 1997). Guha (1995) argues that *Hind Swaraj* is not the right place to look for evidence of Gandhi's environmentalism because he wrote it while living in South Africa, where he spent twenty years as a lawyer and political campaigner.

Many consider the Chipko movement, the NBA, Navdanya, and other Indian environmental movements as living examples of Gandhian environmentalism, and they also consider Gandhi to be a 'man with a deep ecological view of life, a view much too deep even for deep ecology' (Sharma,

2003, p. 45). The Norwegian philosopher Arne Naess, who was the first to coin the term 'deep ecology', posits that ecological preservation is nonviolent in nature. Naess (1988) introduced the relation between nonviolence, self-realization, and mutual dependence of all living beings in the following key points:

(1) Self-realization pre-supposes a search for truth.
(2) All living beings are one.
(3) *Himsa* (violence) against oneself makes self-realization impossible.
(4) *Himsa* against a living being is *himsa* against oneself.
(5) *Himsa* against a living being makes complete self-realization impossible. (Naess 1988, 26)

Naess used these principles to advance deep ecology, a broader philosophy of environmentalism. He believed that Gandhi's utopia is one of the few that shows ecological balance.

5.1.2 Truth at All Costs

Satyagraha, Gandhi's form of holding on to the truth, is an active form of nonviolence. It is based on the idea that a moral appeal to the heart or conscience is more effective than threats of bodily pain or violence. *Satyagraha* originates from the belief that while violence to persons and property diverts the minds of the parties concerned from the real issues at hand, nonviolent action invites the parties to a dialogue about the issues themselves.

It was in the Chipko movement that *satyagraha* was first used as an effective technique against environmental injustice. The forest *satyagraha* of the 1930s were a result of the Forest Act of 1927, which denied villagers access to their forest resources while increasing biomass production for industrial and commercial growth. All of these strands taken together made Gandhi an exponent of Indian environmentalism.

5.2 Defining Views of Globalism

For Gandhi, control of the means of production was central to the development of India after its independence. Likewise, for Navdanya, seeds and land are major means of production leading to the independence of farmers. Just as in the colonial period, when the British government exploited the labour of the peasants to increase productivity, so too, through globalization, do multinationals and transnational corporations deprive peasants of their rights by depleting subsistence agriculture at an accelerated rate.

Although there is no broadly accepted definition of the term 'globalization', the phenomenon has something to do with the way in which the world is changing. Cochrane and Pain (2000) describe it as follows: 'Cultures, economies and politics appear to merge across the globe through the rapid exchange of information, ideas and knowledge, and the investment strategies of global corporations' (cited in Monshipouri et al., p. 968).

Another popular definition of globalization emphasizes that it enhances the level of interaction and interdependence between nation-states and societies. As markets become more intertwined on a worldwide scale, their relationships become deeper politically and economically. On its present neoliberal course, globalization is related to the rationalist structure of knowledge, the capitalist mode of production, technological innovations, and technical and procedural standardization (Scholte, 2000; Monshipouri, Welch, & Kennedy, 2003). Finally, as Monshipouri, Welch, and Kennedy have observed, some scholars see globalization as a paradigm shift in which values, lifestyles, tolerance for diversity, and individual choice simultaneously undergo transformations on a global scale. Globalization as such relates not only to an increased interconnectedness between markets but to a shared culture of globalization, often effecting a social shift away from some traditional ideas and values (Monshipouri & Motameni, 2000; Monshipouri, Welch, & Kennedy, 2003, p. 969).

The language of human rights has tended to reflect this neoliberal globalization. As Monshipouri and Motameni note, the fates of communities throughout the world have become linked through complex and dynamic systems that create moral connections between the agents and the subjects of social action, regardless of territorial and political boundaries. Although there are still varying conceptions of what constitutes rights and what priorities should be assigned to varying types of rights, the philosophical and political debate underlying these disagreements is itself conducted in a global context (Monshipouri, Welch, & Kennedy, 2003).

5.2.1 Technological Prowess

Many developing countries are simply not equipped to participate in this globalized economy on an equal footing. They are instead being plundered by financial capital from the Global North that is accumulating and colonizing their human and natural resources as well as knowledge base developed over centuries. While people once fostered organic relationships and deep attachments to indigenous natural resources, these are now being wastefully used by MNCs. Those who depend on nature and indigenous knowledge for their very subsistence are also the ones who are affected the most by this mindless destruction and wastefulness. With their livelihood taken over by modern mechanized technology and their knowledge base and means of survival rendered redundant or usurped by capital, the poor – especially women – are rendered powerless. The fundamental point that materialist ecofeminism in the Global South 'makes is that the question of the survival of people in the South simply cannot be separated from nature' (Pandey, p. 353).

As in the battle over whether or not to build mega-dams, technology clashes with the desperation to develop economically at the cost of the disadvantaged. This clash is at the centre of worldwide conflicts involving water scarcity, environmental degradation, the loss of biodiversity, globalization, social justice, the survival of indigenous people, and poverty (Leslie, 2005). In the name of economic development, and with the technological means to do it, considerable damage and destruction is being done to common people's lives and environment. As Misra (2007) notes, in the case of the NBA's opposition to the Sardar Sarovar Project, persistent lawsuits, protests, hunger strikes, a review by the World Bank, and a negative impact assessment by an Indian review team did not stop the Indian central government and the state governments of Gujarat and Madhya Pradesh from continuing their funding of the dam.

Pandey (2013) notes that the fact 'That globalization is guided by market forces has meant a complete abandonment of ethics in the actual procedural implementation of neoliberal economics in the countries where it has been introduced' (p. 354). The neoliberal economy, Pandey continues, is dangerously lacking in morals and therefore takes no responsibility for 'the other'. This perverse individualism endangers human security, while from the moral point of view, it demands immediate reparation. This is only possible by introducing ethics into the underpinning philosophy of globalism, alongside the procedural and substantive aspects. 'It is in this context that a codified development ethics and ecofeminism have key roles to play in the revitalization of ethics in the era of globalization' (p. 354).

5.3 Core Values of Development Ethics

Drawing on the works of Gandhi, Freire, Myrdal, Lebret, and others, development ethics emerged as a subfield of development inquiry in the 1970s and

1980s. It is concerned with the evaluation of changes that presume to improve people's lives. Scholars in this field assert the need to seek out, explore, reveal, and test moral values in the prescriptive goals of development theory and practice (Goulet, 1995; Gasper, 2004). In the words of one of the founders of the field, Denis Goulet (1997), 'development ethics has a dual mission: to render the economy more human and to keep hope alive in the face of the seeming impossibility of achieving human development for all' (p. 1161). Two central questions that development ethics seeks to answer are: *What is development?* and *When is it worthwhile?* These questions do not exclude anti-development positions or invasive conceptions of development. They merely demand normative criteria by which to evaluate whatever is regarded as 'development' (Penz et al., 2011).

As part of the policy of 'rapid industrialization' spearheaded by Jawaharlal Nehru, the first prime minister of independent India, massive hydroelectric projects – as well as protests against them – have become part of the social fabric. At the heart of such conflicts are six core values proposed by Penz, Drydyk, and Bose – human well-being and security; equity; participation and empowerment; consideration of human rights; preserving cultures and identities; and environmental sustainability – and the moral issues raised in relation to them, such as the negative outcome of the SSP's Resettlement and Rehabilitation (R&R) policy from the perspective of gender. I will now explore each of these six core values as they relate to the environmental movements depicted in *Displacement by Development: Ethics, Rights and Responsibilities*.

#1 Human Well-Being and Security

With displacement comes a rupture of the social fabric. A key concept related to displacement processes and resettlement programmes is that of well-being, often viewed in material terms. Mehta (2009) points out that 'Women's rights, assets, and spheres of control often centre on informal institutional arrangements, which are rarely captured in policy or understood by policy-makers, and risk being undermined or entirely ignored in the course of resettlement' (p. 7). A process of economic expansion that is damaging to people's well-being – as was true in the cost-benefit analysis made in the Narmada case – is of major concern to critics of the Narmada Valley Development Project (NVDP), since it is not desirable as a social goal. This has important consequences for the meaning of 'development' insofar as the term is used normatively rather than descriptively to advocate processes of social or economic change. Such processes must be conducive rather than damaging to human well-being. As Penz et al. note:

> Therefore, for those who advocate development, well-being becomes a value that is essential to the worth and justifiability of the development enterprise as a whole. Processes of economic expansion or construction that damage or fail to enhance human well-being constitute *maldevelopment* (Penz et al., 2011, p. 121).

The co-founder of the Human Development Index (HDI), Mahbub ul Haq, argues as follows:

> The lack of recognition given to people as an end of development is glaring. Only in the past two decades have we started focusing on who development is for, looking beyond growth in gross national product (GNP). For the first time, we have begun to acknowledge – still with a curious reluctance – that in many societies, GNP can increase while human lives shrivel... We have finally begun to accept the axiom that human welfare – not GNP – is the true end of development. (1995, p. 4)

This conception of development finds support from the moral principle of non-maleficence – which represents the risk side of the risk-benefit analysis – in two ways. Non-maleficence reflects the idea of not inflicting intentional harm but also of not engaging in actions that risk harming others. The first idea relates to development that causes harm, while the second has to do with neglect. Dam infrastructure projects that impoverish the communities they displace are examples of when the principle of minimizing harm has not been considered. A counterargument could be made that greater harm would be inflicted if building infrastructure and industrialization are neglected. But from the standpoint of human well-being, one can argue in retrospect that infrastructure and industrial projects could have been carried out without harming as many people as profoundly. The scope of 'development' that can be advocated in this way must attend to human well-being, at least in saying that processes of economic and infrastructure expansion that cause superfluous harm should be counted as maldevelopment, not as the sort of development that responsible people ought to advocate (Penz et al., 2011).

In the Narmada case, the primary goal has been to increase the availability of hydroelectric power. One might argue that this line of reasoning is sufficient when economic growth was the only goal. But some of the other potential benefits of the dams – particularly flood control and the provision of drinking water – would seem to fit better with the goals of increasing human well-being and security. However, whether such objectives could actually be achieved is questionable. Penz et al. note that flood control has always been a secondary aspect of the original project design, 'while 'quenching the thirst' of drought-stricken regions has so far proven to be a goal fraught with technical and financial problems' (p. 277). It is not only difficult to transport Narmada waters to distant villages in Saurashtra and Kutch, but towns in the industrial heartland of Gujarat have more recently capitalized their own claims to these resources due to their power.

One can also argue that regarding development policies and strategies that do not consider human well-being, fail to do anything for those who are worse off. Penz et al. point out that there is always an alternative of not doing anything but compared to these two alternatives, people who are faithful to the principle of non-maleficence, should favour the policies and strategies which reduces the harm done to the worse off. The other option of tolerating the continuation of avoidable harm, 'cannot be advocated responsibly under the banner of development. To carry out development strategies and policies that preserve and tolerate avoidable harm must therefore, count as maldevelopment' (p. 122). Normatively, therefore, it can be concluded that development must not only avoid processes of economic and societal growth that are harmful to human well-being but also embrace strategies and policies that promote, specifically, the well-being of those who are worse off.

If well-being is an essential aspect of understanding the worth and justifiability of a development enterprise, then the NVDP's record is a problematic one. The earliest justification for the project was a cost-benefit analysis of economic growth that has been discredited by some and defended by others in the development community. One line of reasoning is that to not proceed with the NVDP would threaten the well-being of people in Saurashtra, Kutch, Rajasthan, and a range of other areas that would potentially benefit from the dam. This has been a compelling argument from a political point of view. Penz et al. point out that this argument is deeply troubling because it sets up an opposition between the displaced in the Narmada Valley and the drought-stricken communities in western Gujarat, for example. By setting up this dilemma in this way – that is, choosing the displaced people's side or the drought-stricken areas in western Gujarat – the project raises the issue of neglect or maldevelopment of one side over the other. Why, then, is the NVDP the only viable alternative?

It is important to note here, as Penz et al. point out that many local organizations have argued for solutions to the drought crisis that are based in the villages themselves, with alternatives extending from traditional rainwater harvesting methods and the development of desalination plants to effective gravity-well systems of irrigation. The government of Gujarat has funded very little for these projects, as most of the budget is dedicated to the NVDP. In addition, it is doubtful whether water from the Narmada River would in fact ever reach the drought-stricken area.

#2 *Equity*

For critics of NVDP equity has been a deep-rooted concern. Equity has to do with the relationships among people as related to their circumstances. As Penz et al. suggest, there are three different conceptions of inequity: the unequal sharing of benefits, the imposition or preservation of hardships, and unjust social inequalities. With regard to the first, the NVDP's plans seem to indicate a slanting of benefits towards certain groups at the expense of others. While the political rhetoric has often framed equity issues as the Narmada oustees versus drought-stricken villagers, in terms of actual costs borne and benefits received, it is actually the 'oustees' who endure the cost of displacement. Those who are in the position to benefit – that is, the urban and rural elites – are more likely to gain from the project.

With respect to the imposition or preservation of hardships, the oustees face the cultural and psychological risks accompanying the process, in addition to being physically displaced without appropriate compensation and rehabilitation. There are clearly other groups and individuals than those in Saurashtra and Kutch who will benefit from the NVDP – namely, those located in towns and villages that will receive electricity and gain advantage from increased irrigation. Nevertheless, these benefits must be weighed against the considerable harm to the well-being of those being displaced (Penz et al., 2011).

As for unjust social inequalities, the project transgresses basic notions of social justice: not only are oustees unlikely to become better off through the project, but the record indicates they stand a good chance of becoming even more impoverished and vulnerable. The evidence from existing NVDP dams, such as the Bargi, has shown that the displaced are far from resettled and rehabilitated. Instead, many families and individuals have ended up in the slums of cities such as Jabalpur in Madhya Pradesh, working in the informal sector and maintaining the most precarious of lives. For landless labourers and peasants who have been displaced, one might argue that their plight may not have been made worse, but it certainly has not been made any better either. Their situation involves a violation of John Rawls' egalitarian principle of justice, which argues that unequal treatment can be justified *if and only if* it is necessary, and nobody is made worse off than they are (Penz et al., 2011).

The Narmada case demonstrates the severe danger of perpetuating long- established social inequalities in terms of class, ethnicity, caste, and gender in the project-affected areas (Penz et al., 2011). As an illustration, Mehta (2009) takes the case of an anonymous village activist who was a victim of human rights abuses. A dynamic local leader had challenged gender stereotypes within her family and village as well as amongst the resettlement officials and extension workers who were keen to evict people in the early 1990s. The village, situated in the submergence zone of the SSP in Gujarat, was one of the last areas where some families continued to fiercely resist the project and its associated evictions. To intimidate the community, the police, in collusion with some local leaders, came into the village late at night and arrested the men in the activist's household on false charges and raped the woman. Mehta says, 'After a long enquiry, and despite attempts on the part of the activists to seek justice, the case was dropped due to lack of evidence. The damage to this activist's self-esteem was tremendous, and she became the victim of community-based gender biases' (p. 26).

Many examples of oustee relocation have shown that higher caste groups received preferential treatment in terms of new lands and housing, thereby recreating their existing caste privileges. Resettlement strategies that focus on cash payouts may also increase alienation and insecurity for communities that are used to non-cash economies. As shown above, in light of the unequal sharing of benefits, the imposition or preservation of hardships, and unjust social inequalities, the Narmada case is a clear example of ethical wrongdoing in terms of equity in its three specific dimensions (Penz et al., 2011).

#3 *Participation and Empowerment*

The dams on the Narmada River stand out due to their high social costs, which have been made known to millions around the globe by the activities of the NBA. In addition to calling attention to the dark side of top-down projects such as large dams, the NBA has raised questions about issues that are important for India's future, such as sustainable development; participation of the displaced people in the decision-making process; the rights of indigenous peoples; the viability or non-viability of large, top-down centralist projects; and the mobilization of protest. Due to the protests in the Narmada Valley, the central Indian government and state governments have come up with policies and programmes that, at least on paper, seek to restore and improve the lives and livelihood of the displaced

(Mehta & Srinivasan, 2000). For example, when it was devised in the late 1980s, the Gujarat policy was supposed to be a watershed in government thinking with regard to resettlement and rehabilitation. It provided two hectares (4.94 acres) of irrigable land to all heads of households, including the so-called landless and the encroachers. So-called major sons (above 18 years of age) were also eligible to receive two hectares of irrigable land. In addition, there were provisions for a subsistence allowance and assistance in terms of housing and employment. Mehta shows that though the intention was good, yet these policies were not implemented and had serious flaws from the gender point of view, due to male biases in displacement and R & R processes and didn't take into consideration 'how risks are borne differently by different groups', and how taking care of the risks of some groups 'increase the vulnerability of other weaker and marginalized groups' (Mehta, 2009, p. 8).

Those who have worked for the rights of the displaced have fought for their voices to be heard in other ways as well: by negotiating the types and amounts of compensation that displaced people are entitled to receive; by participating in review boards for impact assessments; and by serving as commissioners involved with the planning of the new elements of the amended NVDP scheme. As Penz et al. says, 'Yet to this day, it is difficult to persuade dam authorities to agree to such participation; indeed, information about the NVDP remains closely guarded, with the most 'sensitive' data protected by India's Official Secrets Act' (p. 281). When the displaced people are given access to information, it is often in a limited form, and their power to act on that knowledge is severely restricted. NGOs and local leaders who seek to work with development planners are often allowed to participate in decision-making, but only partially. 'They are encouraged to aid in dealing with the effects of displacement, but there is little opportunity to participate in macro- development processes' (p. 282).

In response to this limitation, a major division took place with regard to ideology amongst groups protesting the NVDP in 1989. The mostly Gujarati NGOs, which had negotiated for better compensation packages with the government of Gujarat, broke away from the NGOs based in Maharashtra and Madhya Pradesh, which had made out more poorly. It is the latter that would go on and form the NBA. A lack of opportunities for meaningful participation meant that they would not collaborate with the dam authorities, since to do so would have meant to simply validate an illegitimate and unethical project. Participation and empowerment in the Narmada case, therefore, had been wholly unsatisfactory (Penz et al., 2011).

#4 *Consideration of Human Rights*

Environmentalists and human rights activists point out that mega-dams displace millions of people and have crippling effects on 'riverine ecosystems, floodplains, woodlands, wildlife, fisheries, and forests that local communities rely on to protect their livelihoods' (Agoramoorthy et al., 2008, p. 26). India's Sardar Sarovar Dam became one of the world's most controversial dam projects due to the negative effect of the displacement of people. Most of the affected people were poor indigenous tribal communities. Several environmental and anti-dam groups, especially the NBA, continue to advocate for those who have been affected, bringing hundreds of legal cases against the government to fight for the rights of those displaced (Agoramoorthy et al., 2008). The NBA has argued that the NVDP does not respect the rights of indigenous groups to be resettled as communities, nor has it ensured that oustees' political, social, economic, and cultural rights are protected. In terms of its neglect of human rights, therefore, the Narmada Valley Development Project may be viewed as maldevelopment (Penz et al., 2011).

Mehta (2008) notes that the Narmada Project will have an adverse effect on the homes, lands, and livelihood of about a million people. About 250,000 people will be directly impacted and will lose their homes due to reservoir submergence in Gujarat, Maharashtra, and Madhya Pradesh. The rest will be indirectly affected due to the effects of the canal construction and downstream impacts of the dam. A high percentage of this population is *adivasi* (tribal populations).

It seems clear that as a result of the risks that they may suffer, the oustees from the Narmada Valley face the possibility of an infringement of their human rights. The issue has come to the forefront clearly in the NVDP case due to the involvement of a large number of indigenous and marginalized people. In this context, it is not merely that potential oustees are vulnerable but they are also 'entitled to a series of ethical and legal protections under both international and national law', while the actual laws in place do not give much attention to what is ethically required (Penz et al., 2011, p. 283).

#5 *Preserving Cultures and Identities*

The intertwining of local livelihood with spiritual, if not religious, meanings was seen in many other expressions of environmental protests across India. Two of the most dramatic and influential movements were related to the displacement and environmental destruction caused by the construction of big dams in India: the anti-dam movements of Tehri Garhwal (also in the Himalayan Ganges region) in the 1980s, and the protest against the Narmada River dam (Sardar Sarovar Dam) in Gujarat during the 1990s. As a result of these protest movements and their alliances with other large dam protests across the world, the Indian state has largely stopped its construction of large dams, and the World Bank and other transnational agencies have reversed the trend of funding large dam projects (Duara, 2015).

The Narmada controversy is not based on a simple disagreement about whether dam-building is viable or not. There are philosophical, political, and moral implications involved in relation to the contemporary model of development (Misra, 2007). The Narmada River valley is blanketed with numerous shrines, temples, and cultural artifacts. This is highly significant to the majority of oustees and the religious perspective of Hindus. The loss of these cultural treasures cannot be easily offset with monetary compensation (Penz et al., 2011). As Misra (2007) claims, the Indian state and development agencies like the World Bank seem to be disregarding the historical, cultural, and spiritual ties that people have to the land. They view the land as a mere commodity (Basu & Silliman, 2000). The Narmada movement is opposed to this limited view of land as a commodity. For them, the land is a living thing with many components. The question becomes whether the destruction of such sites and objects in the course of the dam-building can be justified. The NVDP has had serious deleterious effects upon indigenous and otherwise culturally unique populations (Misra, 2007).

The religious and cultural ties to the river, very prominent in India, are often stressed by activists. In Hindu and tribal cultures, the Narmada river is viewed as *mata* (mother), and maternal symbolism is an integral part of the anti-dam resistance. The submergence of holy pilgrimage sites and sacred monuments along the Narmada Valley is also a major concern of the tribal people. The image of Gandhi and the use of Gandhian strategies of nonviolence are striking (Fisher, 2000). The NBA achieved large-scale national and international support because of its unique orientation and strategies of resistance (Misra, 2007). The NBA followed the Gandhian method of *satyagraha* by their commitment to the principle of justice without using any form of violence, and at the same time protesting to preserve the sacred monuments that are essential to the identity and culture of the Narmada Valley (Misra, 2007).

#6 Environmental Sustainability

In the Narmada case, the issues of ecology and sustainability are perplexing. On the one hand, the environmental impacts are serious and difficult to remedy. On the other hand, for a growing, industrial economy and society such as India, with its uncontrollable appetite for energy, hydroelectricity is seemingly a far more palatable source than alternatives such as coal or nuclear power. Indeed, this is the argument advocated by proponents of the dam. As Penz et al. state, 'In recent years, however, the viability of hydroelectric power as a form of 'green' energy has been called into question. Critics point to the direct impact of dams such as the loss of forests and fauna in the flooding to create the reservoirs, the disruption of river systems and river flows, and potential tectonic risks from the weight of the reservoir load' (p. 285). Moreover, as global warming and climate change become ever more prominent scientific and political concerns, large dams have been questioned in terms of their contribution to carbon and methane emissions from their vast reservoirs.

Indeed, it is precisely on grounds of environmental sustainability that the NVDP has met the greatest resistance, outside of popular protests and interstate disputes. Penz et al. note that it was due to environmental reviews by the Ministry of Environment and Forests (MoEF) that construction in the 1980s was delayed; and the World Bank used the lack of ecological safeguards as one important justification for withdrawing its funding for the project. Lastly, it was precisely on grounds of environmental harm that the Supreme Court suspended the NVDP in the 1990s (2011). This suspension was, however, later lifted.

5.3.1 Environmental Justice for All

The NVDP raises many moral issues in relation to the six core moral values advocated by the proponents of development ethics. The core values need to be addressed in order to arrive at a better situation for those who have been negatively impacted by the dam. But how are gender issues mirrored in the resettlement and rehabilitation of the displaced people?

The global environmental justice movement started as a political and social manifestation of growing concern over the inequitable distribution of hazardous industries and their waste among poor and minority communities (Bullard, 1990, 1993). Soon after its inception, the movement also addressed the need for social justice issues to be incorporated as central dimensions in traditional environmentalism (Jamieson, 2007) in ways that honour diverse cultural perspectives (Di Chiro, 1998).

Environmental justice is gendered because women suffer most from poverty, human rights violations, and environmental destruction (Buckingham et al., 2005). Although a number of campaigns have signaled the disproportionate impacts of environmental problems on women, the focus on gender inequalities has yet to become a prominent or widespread feature in either environmental justice campaigns or academic analyses of such campaigns (Buckingham et al., 2005). Buckingham and Kulcur (2009) examine some of the reasons why this may be so, mentioning four possible factors:

(1) lack of visibility of gendered environmental injustice;
(2) professional campaigning organizations which are themselves gender-blind;
(3) institutions that are still structured by gender as well as by class and race, reflecting inequalities at a range of scales;
(4) an intellectual academy that continues to marginalize the study of gender and women's inequality. (p. 661)

5.4 Ecofeminism: Ethics of Mutual Care and Connection

Ecofeminism shares some of the characteristics of the worldwide environmental justice movement, because women are more likely than men to suffer from poverty. Moreover, women predominate in environmental justice grassroots movements, although gender has not been associated with the environmental justice field (Buckingham & Kulcur, 2009; Di Chiro, 1998; Taylor, 2000; Unger, 2008; Sakellari & Skanavis, 2013). Pandey (2013) notes that while the West makes an issue of reclaiming or waiving developing countries' debt, ecofeminists such as Salleh (2009) highlight the largely ignored problem of 'ecological and embodied debt'. The concept is based on the argument that wealthy developed nations owe reparations to developing countries due to the simple fact that the former were built on the depletion of the ecological resources and labour power of the latter. This is indeed the central thesis of ecofeminist Maria Mies (1986), who argues that the developed countries have been built on the backs of nature, women, and the colonies.

Pandey notes that ecofeminism offers a choice – an alternative way of life that situates itself within the organic whole and finds its fulfillment through a spiritual and material union with nature. Ecofeminism offers a world where prosperity is built not on the exploitation of the weak and voiceless but on caring, nurturing, and symbiotic relationships instead. Apart from the urgent considerations of planetary survival and well-being, the primary reason for adopting an ecofeminist approach is the desperate need for ethics to examine carefully how we ought to relate to each other in forming relationships based on differences of gender, race, national origin, and species. It is precisely overcoming the self/other dualism that is key to the construction of an ecofeminist ethics (Plumwood, 2002, 2000, 1993). The self/other dichotomy is a reflection of the deeper dualism of reason/nature wherein reason tends to exclude the other and all that it regards as 'nature'. Dualism, Pandey says, 'takes on an instrumental attitude toward nature which in turn involves hyper separation between the self and the other so as to highlight differences rather than acknowledge the similarities or the bonds and connections between them' (pp. 353-354).

In this age of globalization, it is paramount that we find an alternative to the current relationship of absolute domination and hierarchy between those who stand to gain the most from globalization and those on whose back prosperity is created. What we have, essentially, is a minority of the world's population and nations – specifically the white, male-dominated population of the developed world (and also women, though to a lesser extent) – exploiting the natural resources and people of the former colonies in the developing and least-developed countries. It is in this context that ecofeminism makes an invaluable contribution to the relationship between the self and the other by postulating a relationship of symbiosis, trust, and care without mistaking one's identity with that of the other. In the words of Plumwood (2002): 'We need a concept of the other as interconnected with self, but as also a separate being in their own right, accepting the "uncontrollable tenacious otherness" of the world as a condition of freedom and identity for both self and the other' (p. 201).

Thus, Plumwood emphasizes the fact that it is possible to display a sense of political solidarity with the other – in this case, nature – without dissolving one's identity with that of the other. As Mallory (2009) explains, 'For Plumwood, solidarity entails recognizing and accounting for the difference of the Earth other while not avoiding our responsibility to act politically on its behalf' (p. 13). Pandey (2013) argues that the relationship between nature and women in the developing countries is even more complex and nuanced than Plumwood recognizes. Pandey notes, 'Although it is very much based on the respect for 'difference' that Plumwood espoused', this relationship involves 'a deeper sense of unity and bonding', since 'it is impossible to separate the spiritual identity of the woman agent from her material bases' in developing countries (Mies & Shiva, 1993; Sydee & Beder, 2001; also cited in Pandey, p. 353).

In an ecofeminist ethic, the concept of possession and ownership – as in property ownership in capitalism – is replaced by a sense of belonging. As

Stephens (2009) notes: 'Belonging is the bond of love and respect that grows in the course of the seasons' (p. 64). The support and bond come from care rather than the desire for power. Stephens denounces the zero-sum game of continuous accumulation, which tends to be based on power.

5.4.1 Ecofeminist Roots in the Chipko Movement

The actions of the women involved in the Chipko movement are eloquently described by Vandana Shiva as 'the ultimate connection with and love for nature, to the point that they were willing to sacrifice their lives' (Mauro YouTube Video, 2013). Therefore, even though the Chipko movement is well-known today for being one of the first successful, nonviolent environmental protests, it holds a deeper meaning and significance for ecofeminists as the first political and environmental movement led by traditionally subordinated women fighting for recognition by their state and their communities. Female environmental activists have for a very long period played a crucial role in creating ecological awareness and pushing for environmental protection (Zelezny & Bailey, 2006).

Although the Chipko movement was not a politically feminist movement as such, the women who participated in meetings, processes, and other programmes became examples for subordinated women in India and elsewhere, by showing their ability to participate in the political process as a response to the immoral behaviour of the people in power (Li, 2013). Women gave strength through their direct support from the beginning of the movement until the end, commendable as it is one of the first demonstrations of ecofeminism in action. As feminist Mary Mellor (2008) shows, by fighting for their rights to their land and natural resources, Chipko women became the voice of women around the world who faced similar circumstances of suppression and political exclusion in their societies. Thus, the women of the Chipko movement can be said to represent a situated yet universal perspective in the words of Karen Warren's later idea (1987, 1990) that paved the way for ecofeminism. This has demonstrated that women often face the same societal subordination and therefore respond in a way that represents all women in similar situations, regardless of time or location.

In order to understand women's devotion to the Chipko movement, as Li states in her article, 'The Roots of Ecofeminism: The Women of the Chipko Movement' (2013), it is necessary to understand the social setting of the local villages and the inferior gendered roles of women in these settings. Jain's case study (1984) points out that egalitarianism did not exist in homes and family situations. The community structures, power, and authority were built around a hierarchy of males. Men were generally in charge of farming; women's roles in the home and community were to be caretakers of children and to collect firewood and fetch water for the essential needs of the family (Jain, 1984). Because their jobs involved collecting firewood, women were the first to instinctively realize the extreme effects of commercial logging on their environment and livelihood. Moreover, the women realized that the way they had been treated and taken advantage of by their community was immoral.

Jain highlights the point that despite the oppression faced by women from their male counterparts and their deteriorating status in their society, their tree-hugging activism showed that they were not only capable of making decisions but also able to take care of their survival without male leadership. Vandana Shiva was a Chipko activist herself. She states in an interview with Ian Mauro that without the quick thinking of these self-sacrificing women, the Chipko movement would not have achieved the victory it did, with the Indian government finally issuing a ban on the felling of trees in the Himalayan regions for fifteen years until the green cover was fully restored (Mauro, 2013).

In this way, Chipko has exemplified the ecofeminist ideas of Warren, Mellor, and others. Ecofeminists see gender inequality as the result of a destructive relationship between humanity and the natural world. As the domination of women and nature is fundamentally connected, women's environmental efforts are integral to overcoming the oppression by themselves. Ecofeminism emphasizes that women do not necessarily seek equality with men but aim at the liberation of women by recognizing their indispensable role in society, having objectives similar to nature (The Green Fuse, 2012).

The Chipko movement has impacted ecofeminist debates around the social and economic factors that dictate women's inherent connection with nature in several ways. First, ecofeminists emphasize that women have a physiological connection with nature due to their social experiences with birth and childcare, which largely ties back to the same gendered roles of tribal women in the Garhwal Himalayas. Roucheleau et al. (1996) point out that while our modern societies might not be as patriarchal today, women still hold a traditionally inferior role to men in the household and in society; she explains women's close relationship with nature as 'patriarchal dualism'.

Second, Mellor (2008) argues that women's social role is also shaped by their economic role in the sense that women's subordination lies in their centrality to the 'support economies' of unpaid domestic work in the home and the community (pp. 69-70). This is significant, because unlike the labour in male-dominated and profit-driven economies, a woman's work in the household is inevitably very responsive to environmental factors such as the health of the local ecosystem. This was especially seen in subsistence economies such as the Garhwal Himalayas, where women's jobs were critically embedded in the local environment. Thus, ecofeminists believe that women are instinctively much more socially and economically concerned with the environment because they see how their basic subsistence and health is dependent upon it, regardless of economic incentives. This concern was evident when the women of the Chipko movement recognized the effects of deforestation before any of the men, who were more concerned with losing their crops than the environmental issues at hand.

It is important to note that the Chipko women represented not only their particular cause to stop commercial logging and destruction of their homes but also women around the world who suffer from the same political neglect and societal suppression in typically male-dominated societies. Thus, the bold and brave women's activism in the Chipko movement to stand up for themselves and their environment shaped not only their own fate in society but the ideals and principles of the modern-day ecofeminist movement against the patriarchal dualism of women and nature in all societies. Culturally situated and yet universally dispossessed, the Chipko movement's connectedness to nature and steadfast political and social activism was an embodied reaction to traditional gendered roles and their subordination to men in a patriarchal society (Li, 2013).

5.4.2 Southern Materialist Ecofeminism

In Pandey's words, 'A materialist ecofeminism is premised on its ability to care for the other, based on taking due cognizance of existing relationships and bonds between the carer and the caree and recognizes individuals as real, concrete individuals embroiled in a network of relationships' (p. 354). It promotes a way of life that boldly acknowledges dependence on the other and is not afraid to assume responsibility for the same. On the contrary, the globalized economy sees nature as well as human lives as nothing more than a resource to be exploited, used, or at best 'managed'. An ecofeminist ethics of care primarily rejects the predisposition to use or exploit the other. It broadens the scope of human social relations by situating them within the context of nature. It is premised on the need to derive ethics from ties, bonds, attachments, and relationships that grow out of the moral agent's relationships to others around herself. A materialist ecofeminism is grounded in the recognition of the fact that human beings are both social and natural beings. Hence, it aims to situate human beings within a specific context of relationships within society and nature (Pandey, 2013).

In fulfilling one's moral responsibility toward nature, ecofeminists are only recognizing all that makes people human, that is, their relationships with others within a complex web of society and ecology. It is through these bonds that we formulate principles of ethics in the first place. In realizing their ethical responsibilities toward the interests of human and 'non-human' others, ecofeminists are securing their own material and emotional needs (Pandey, 2013). The critical point that materialist ecofeminism is making with respect to globalization is that not anthropocentric practices, but practices of certain humans are the cause of environmental degradation. Cuomo (1998) names corporate greed, militarism, nationalism, and ethnocentrism among such questionable practices. In addition, Pandey (2013) characterizes capitalist globalization and the processes associated with it as the culprits most responsible for jeopardizing nature and human existence, because social justice cannot be separated from ecological issues.

5.5 Conclusion

This chapter has conveyed one of the main messages of this book, which is that the ethical aspect of the three movements cannot be ignored. In making decisions, the value and well-being of all human beings – including the marginalized population – cannot be subordinate to economic, scientific, and technological growth. The disadvantaged in society who are negatively affected by deforestation, dam-building, and GMOs must be given an equal voice and cannot be considered less important than the interests of the powerful, the state, and the major corporations.

This chapter has analyzed the significance of nonviolence as a method of active resistance in the Gandhian context. In the debate over how Gandhi has inspired environmentalists, some scholars have argued that Gandhi was not an environmentalist, as there is simply not enough textual evidence in Gandhi's writing to claim Gandhi as the father of environmentalism in India. After surveying some of the literature on globalism, I have argued that neoliberal globalization does not lead to a level playing field for the Global South. In the Southern Hemisphere, people's survival is intricately related to nature. Neoliberal globalization loses sight of the moral aspect when, as a matter of fact, it is essential to take into consideration the moral aspect and repair the wrongs that have been done to the poor and the marginalized indigenous population.

In this regard, it is significant that the emerging field of development ethics pays attention to human development for all rather than a certain section of the population. By applying the six core values of development ethics as proposed by Penz et al., I have demonstrated that there are many moral issues affecting the groups that have been displaced by dam-building, and these are issues that require close attention and consideration. These core values demand that normative criteria be used to evaluate any kind of development. The powerlessness of the displaced people and the harm that has been done have been significant and complex.

I also showcased the relation between gender and the Narmada case by arguing that the Resettlement and Rehabilitation (R&R) policies are not gender-blind. Gender ought to be an essential element of such R&R policies. I established the connection between economic justice and ecofeminism by showing how ecofeminism explores the relation between the self and the other. The Chipko movement can be related to ecofeminism due to the activist role played by women, which has influenced women not only in India but also in other parts of the world. Materialist ecofeminism originating from the Global South emphasizes an ethics of care, arguing that social justice and ecological issues are related to each other. This ethics of care rejects the exploitation of 'the other', whereby 'the other' is extended to include not only other humans but also nature.

The next chapter addresses the subject of Hindu ethics and the role it plays in the environmental movements of India through an analysis of several Hindu scriptures and what they have to say about the interconnectedness of humans, nonhumans, and nature.

Works Cited

Agoramoorthy, Govindasamy, and Minna J. Hsu (2008). 'Small Size, Big Potential: Check Dams for Sustainable Development'. *Environment Science and Policy for Sustainable Development*, 50, 22-35.

Basu, Pratyusha, and Jael Silliman (2000). 'Green and Red Not Saffron: Gender and the Politics of Resistance in Narmada Valley'. In Christopher Key Chapple and Mary Evelyn Tucker (eds.), *Hinduism and Ecology: The Intersection of Earth, Sky and Water.* Cambridge, MA: Harvard University Press, 423-452.

Buckingham, Susan, Dory Reeves, and Anna Batchelor (2005). 'Wasting Women: The Environmental Justice of Including Women in Municipal Waste Management'. *Local Environment*, 10, 4, 427-444.

Buckingham, Susan, and Rakibe Kulcur (2009). 'Gendered Geographies of Environmental Injustice'. *Antipode*, 41, 4, 659-683.

Bullard, Robert D. (1990). *Dumping in Dixie: Race, Class, and Environmental Quality.* Westview Press.

— (1993). 'Anatomy of environmental racism and the environmental justice movement'. In R.D. Bullard (ed.), *Confronting Environmental Racism: Voices from the Grassroots.* Boston, MA: South End Press.

Cochrane, Allan, and Kathy Pain (2000). 'A Globalizing Society'. In David Held (ed.), *A Globalizing World? Culture, Economics, Politics.* Psychology Press, 5-21.

Cuomo, Chris (1998). *Feminism and Ecological Communities: An Ethics of Flourishing.* London: Routledge.

Di Chiro, Giovanna (1998). 'Environmental Justice from the Grassroots: Reflections on History, Gender and Expertise'. In Daniel Richard Faber (ed.), *The Struggle for Ecological Democracy: Environmental Justice Movements in the United States.* New York: Guilford Press, 104-136.

Duara, Prasenjit (2015). *The Crisis of Global Modernity.* Cambridge: Cambridge University Press.

Fisher, William F. (2000). 'Sacred Rivers, Sacred Dams: Competing Visions of Social Justice and Sustainable Development along the Narmada'. In Christopher Key Chapple and Mariy Evelyn Tucker (eds.), *Hinduism and Ecology: The Intersection of Earth, Sky and Water.* Cambridge, MA: Harvard University Press, 410-423.

Gandhi, Mohandas Karamchand (1909). *Hind Swaraj,* Gandhi Heritage Portal.

— (1931). *Young India 1919-1922.* New York: B.W. Huebsch.

— (1936). *Harijan,* September 5, Gandhi Heritage Portal.

Gasper, Desmond (2004). *The Ethics of Development.* Edinburgh University Press.

Godrej, Farah (2012). 'Ascetics, Warriors and a Gandhian Ecological Citizenship'. *Political Theory,* 40, 4, 437-465.

Goulet, Denis (1995). *Development Ethics: A Guide to Theory and Practice.* New York: Apex Press.

— (1997). 'A New Discipline: Development Ethics'. *International Journal of Social Economics,* 24, 1160-1171.

Guha, Ramachandra (1995). 'Mahatma Gandhi and the Environmental Movement of India'. *Capitalism, Nature, Socialism,* 6, 3, 47-61.

—, and Joan Martínez-Alier (1997). *Varieties of Environmentalism: Essays North and South.* London: Earthscan Publications.

Haq, Mahboub ul (1995). *Reflections in Human Development.* New York: Oxford University Press.

Jain, Shobhita (1984). 'Women and People's Ecological Movement: A Case Study of Women's Role in the Chipko Movement in Uttar Pradesh'. *Economic and Political Weekly,* 19, 41, 1788-1794.

Jamieson, Dale (2007). 'Justice: The Heart of Environmentalism'. In Ronald Sandler and Phaedra C. Pezzullo (eds.), *Environmental Justice and Environmentalism: The Social Justice Challenge to the Environmental Movement*. Cambridge, MA: The MIT Press, 85-101.

Kumar, Devendra (1992). 'Kumarappa and the Contemporary Development Perspective'. *Gandhi Marg*, 14, 294-315.

Leslie, Jacques (2005). *Deep Water: The Epic Struggle over Dams, Displaced People, and the Environment.* New York: Farrar, Straus and Giroux.

Li, Huey- li (2007). 'Ecofeminism as Pedagogical Project: Women, Nature and Education'. *Educational Theory*, 57, 3, 351-368.

Li, Wendy (2013). 'The Roots of Ecofeminism: The Women of the Chipko Movement'. *Social Science News*, permission of author.

Mallory, Choane (2009). 'Val Plumwood and Ecofeminist Solidarity: Standing with the Natural Other'. *Ethics and the Environment*, 14, 2, 3-21.

Mauro, Ian (2013). 'A Conversation with Vandana Shiva – Question 3 – Tree-hugging and the Chipko Movement'. Mount Allison University. YouTube Video.

Mehta, Lyla (2008). 'Why are Human Rights Violated with Impunity? Forced Displacement in India's Narmada Valley'. In Katarzyna Grabska and Lyla Mehta (eds.), *Forced Displacement: Why Rights Matter.* New York: Palgrave Macmillan, 201-221.

— (2009). 'The Double Bind: A Gender Analysis of Forced Displacement and Resettlement'. In Lyla Mehta (ed.), *Displaced by Development: Confronting Marginalization and Gender Injustice*. Sage Publications, 3-33.

—, and Srinivasan, Bina (2000). 'Balancing Pains and Gains: A Perspective Paper on Gender and Large Dams' (Report prepared for the World Commission on Dams Thematic Review: Social Impacts).

Mellor, Mary (2008). 'Ecofeminism: Linking Gender and Ecology'. In T. Jules Pretty et al. (eds.), *The Sage Handbook of Environment and Society.* Sage Publications, 66-77.

Mies, Maria (1986). *Patriarchy and Capitalist Accumulation on a World Scale.* London: Zed Books.

—, and Vandana Shiva (1993). *Ecofeminism.* London: Zed Books.

Misra, Shalini (2007). 'Spirituality, Culture and the Politics of Environmentalism in India'. *Journal of Entrepreneurship*, 16, 131-145.

Monshipouri, Mahmood, and Reza Motameni (2000). 'Globalization, Sacred Beliefs, and Defiance: Is Human Rights Discourse Relevant in the Muslim World'? *Journal of Church and State*, 709-736.

Monshipouri, Mahmood, Claude E. Welch, Jr., and Evan T. Kennedy (2003). 'Multinational Corporations and the Ethics of Global Responsibility: Problems and Possibilities'. *Human Rights Quarterly*, 25, 965-989.

Naess, Arne (1988). 'Self-Realization: An Ecological Approach to Being in the World'. In John Seed and Joanna Macy (eds.), *Thinking Like a Mountain: Towards a Council of All Beings.* Philadelphia: Society Publishers, 19-31.

Pandey, Anupam (2013). 'Globalization and Ecofeminism in the South: Keeping the "Third World" Alive'. *Journal of Global Ethics,* 9, 3, 345-358.

Penz, Peter, Jay Drydyk, and Pablo S. Bose (2011). *Displacement by Development: Ethics, Rights and Responsibilities.* Cambridge: Cambridge University Press.

Plumwood, Val (1993). *Feminism and Mastery of Nature.* London: Routledge.

—(2000). 'Integrating Ethical Frameworks for Animals, Humans and Nature: A Critical Feminist Eco-Socialist Analysis'. *Ethics and the Environment,* 5, 2, 285-322.

—(2002). *Environmental Culture: The Ecological Crisis of Reason.* London: Routledge.

Roucheleau, Dianne, Barbara Thomas-Slayter, and Esther Wangari (1996). 'Gender and Environment: A Feminist Political Ecology Perspective'. In Nora Haenn and Richard R. Wilk (eds.), Reprint Edition, (2016). *The Environment in Anthropology: A Reader in Ecology, Culture and Sustainable Living.* New York: New York University Press, 34-40.

Sakellari, Maria, and Constantina Skanavis (2013). 'Environmental Behaviour and Gender: An Emerging Area of Concern for Environmental Education Research'. *Applied Environmental Education and Communication,* 12, 77-87.

Salleh, Ariel (2009). *Eco-sufficiency and Global Justice: Women Write Political Ecology.* London: Pluto Press.

Scholte, Jan Aart (2000). *Globalization: A Critical Introduction.* London and New York: Palgrave.

Sharma, R.C. (2003). *Gandhian Environmentalism.* New Delhi: Global Vision.

Sydee, Jasmine, and Sharon Beder (2001). 'Ecofeminism and Globalization: A Critical Appraisal'. *Democracy and Nature,* 7, 2, 281-302.

Stephens, Piers H.G. (2009). 'Plumwood, Property, Selfhood and Sustainability'. *Ethics and the Environment,* 14, 2, 57-73.

Taylor, Dorceta E. (2000). 'The Rise of the Environmental Justice Paradigm: Injustice Framing and the Social Construction of Environmental Discourses'. *American Behavioural Scientist,* 43, 4, 508-580.

Terchek, Ronald J. (2011). 'Conflict and Nonviolence'. In Judith M. Brown, and Anthony Parel (eds.), *A Cambridge Companion to Gandhi.* Cambridge: Cambridge University Press, 117-134.

The Green Fuse, Ecofeminism (2012). Electronic document http://www.thegreenfuse.org/ecofem.htm (accessed 8 December 2018).

Unger, Nancy (2008). 'The Role of Gender in Environmental Justice'. *Environmental Justice,* 1, 3, 115-120.

Warren, Karen J. (1987). 'Feminism and Ecology: Making Connections'. *Environmental Ethics,* 9, 1, 3-20.

— (1990). 'The Power and The Promise of Ecological Feminism'. *Environmental Ethics,* 12, 2, 125-146.

Weber, Thomas (1999). 'Gandhi, Deep Ecology, Peace Research and Buddhist Economics'. *Journal of Peace Research,* 36, 3, 349-361.

Zelezny, Lynnette, and Megan Bailey (2006). 'A Call for Women to Lead a Different Environmental Movement'. *Organization and Environment,* 19, 1, 103-109.

6 Hindu Ethics and Ecology

Abstract
This chapter provides a detailed analysis of Hinduism, examining some popular Hindu scriptures in which the interconnectedness of humans and nature are depicted and noting how these scriptures are used only selectively by the general population of India without much attention being paid to the environmental crisis. The analysis demonstrates that Hinduism has a metaphysical basis for ethical guidance and that Hindu ethics based on *dharma* with its many levels – the universal and the particular (based on one's status, caste, and family) in contrast to ethics viewed at the universal level in the Western context – can be viewed as the foundation of environmental sustainability. Following an extensive review of the literature, it is argued that Hindu scriptures should be the basis of reducing the environmental damage caused by humans.

Keywords: stages of Hinduism, *dharma*, ethics, religious environmentalism, *Advaita* and *Vishistadvaita Vedanta*

Whether or not the environmental movements of India have their philosophical foundations in the ethics of Hinduism has been open to debate. Some scholars argue that although there are many philosophical ideas proposed in different scriptures of Hinduism, the environmental movements are not based on those philosophical and ethical ideals. They claim that environmental movements are basically concerned with issues related to the livelihood of the people who are directly affected and to how they are treated by others. Other scholars counter that there is a connection between Hindu ethics and sustainability and that Hindu ideals should be reinstated in the sustainability discourse. The focus of this chapter is to try, like Bilimoria (1998), 'to identify patterns of ideas which may complement the history of environmental thinking' (p. 1). By analyzing the pivotal views of experts in the field of religious environmentalism, I will show how the complexities of Hindu ethics have evolved over the centuries and how it

Mallick, Krishna, *Environmental Movements of India: Chipko, Narmada Bachao Andolan, Navdanya*. Taylor & Francis Group, 2021.
DOI: 10.5117/9789462984431_CH06

is related to sustainability in the present context, particularly since the emergence of the interdisciplinary subfield of religious environmentalism in the 1990s. A variety of philosophical and ethical ideas of Hinduism are explored that may relate to sustainability.

Hinduism, unlike Western religions, has a variety of scriptures that are spiritually, socially, culturally, philosophically, and ethically linked. These scriptures have developed over thousands of years and describe the history of demographic movements, the reformations that society has gone through, and the foreign interference that led to changes in structure and identities. Although the scriptures of Hinduism were clearly not written in response to the ecological destruction created by humans in the last two centuries, they do offer a viable solution to the environmental crisis. It can be argued that at the stage of ecological crisis that we are in, entering the third decade of the 21st century, these Hindu scriptures can and should motivate a change in our relationship with nature.

Part of the problem is a disconnect between nature being valued in many religious scriptures of Hinduism and these scriptures being followed only selectively by the general population. This is due to the complexity and evolution of Hinduism, which has different levels from which it has been approached by Hindus and non-Hindus in India. There is a need for a renewed emphasis on the harmony and sacredness of nature as depicted in the Hindu religious scriptures. For this to happen, the general population must be educated to move away from a consumerist mentality and move toward some form of spirituality following a moral code to achieve the broader goals of sustainability and a harmonious co-existence with nature. In Hinduism and in any of the Eastern religions, religion and philosophy are interrelated and not viewed as separate disciplines, as in the Western religions. Philosophy is a way of life in India. It is the quest for meaning that one should live out in one's day-to-day life.

Hinduism is an aggregate of many different traditions; some are theistic, some are affirming of life, and some are ascetic. The *Vedas* (1500-900 B.C.) are considered to be the earliest written scriptures. It is important to note here that before the *Vedas* were published in written form, they were transmitted orally from teachers to students from approximately 2,500 B.C., though historians do not agree on the exact date. The *Upanishads* (800-500 B.C.) developed from the *Vedas*, and in turn the six Indian orthodox philosophical systems – *Samkhya*, *Yoga*, *Nyaya*, *Vaisheshika*, *Mimamsa*, and *Vedanta* – and two heterodox or atheistic systems (*Buddhism* and *Jainism*) developed from the *Upanishads*. It is metaphysics, not ethics, that is the foundation of the schools of Indian philosophy, but there is also a Hindu ethics that serves as guidance for this metaphysical foundation. Hinduism is naturally pluralistic. A well-known hymn from the *Rigveda* says: 'Truth is One, though the sages know it variously' (in Sanskrit, *Ékam sat vipra bahudā vadanti*).

India was not a poor country before British rule. It was a rich country until the British took over India, after which it became a poor country. It has always been rich in thought. In the Western world, philosophy originated from a sense of wonder and curiosity about life, but in the context of India, philosophy came from speculating and reflecting on nature, the functioning of living beings, and the meaning of language. *Darshan* – meaning 'to see' or vision – is the Sanskrit word for philosophy. It also refers to the auspicious sight of a deity or having a vision of truth, and it includes the six orthodox schools of philosophy.

6.1 Historical Background of Hinduism

Hindu religious periods can be divided into five stages. Jhingram (1989) points out that the first stage of Vedic 'religio-culture' tradition 'is rich in mundane values and is generally characterized by a positive or affirmative attitude towards the world and life' (p. 10). This positive attitude became more pronounced later on, when prosperity (*artha*), the fulfillment of desires (*kama*), and righteousness (*dharma*) together were the human goal, as stated in the different Vedic texts. Religious scholars distinguish between the Vedic period (also called Brahmanism) and the emergence of Hinduism in the classical period when the *Upanishads* emerged. The *Vedas* were created by the Aryans, who invaded India from Persia. Brahmanism followed the Vedic tradition of fulfilling one's desire, the performance of rituals. The ancient Aryans also accepted violence as a part of life. According to Jhingram, 'In fact, killing was an accepted duty of *kshatriyas* in the ancient Hindu society, and it was assured that a *kshatriya* who is killed in war goes to heaven' (1989, p. 11).

The Sramana tradition developed side by side with Brahmanism in the second stage. This tradition advocated the ascetic life through self-discipline and renunciation of the married and domestic life. This second stage of the religio-cultural tradition gave birth to the great philosophies of India. Jhingram notes that people were preoccupied with the mystery of life, and there was a rise in intellectual and spiritual unrest that stimulated the Aryans – in particular the *kshatriyas* (the warrior caste) – to develop the philosophical schools of *Jainism*, *Buddhism*, *Samkhya*, and *Vedanta* around 660 B.C. These schools of philosophy rejected the rituals and violence of the Vedic period. Instead, their focus was on achieving liberation or *moksha* by transcending this world. Liberation has been viewed in different ways by these schools, whose philosophers developed their philosophies independently but at the same time interacted with each other.

Buddhism and *Jainism* were heterodox or atheistic schools that developed around 660 B.C. The orthodox system of philosophy known as *Samkhya* developed independently during this time. *Samkhya*, which is an orthodox school, and *Jainism* and *Buddhism*, both of which are atheistic schools, have many ideas in common but are very different from the *Vedanta* school of philosophy.

The *Vedanta* school of philosophy developed between approximately100 and 500 B.C. In its monistic form later known as *Advaita Vedanta*, this school believed that liberation or *moksha* could be achieved through the 'intuitive realization of the metaphysical truth' (Jhingram, 1989, p. 14). In this quest for liberation, ethical discipline and conduct are considered a prerequisite, though they are not directly instrumental to the achievement of liberation.

Samkhya believes in two ultimate realities: *purusha* (self) and *prakriti* (nature). It believes in the plurality of selves, transmigratory existence, and the essential duality between the self and nature. Liberation according to the *Samkhya* school of philosophy is the self's realization of its original being. *Vedanta*, by contrast, affirms the all-comprehending Absolute or *Brahman*, the reality or Self over all individual selves. The *Buddhist-Jaina-Samkhya* schools perceived the world as full of suffering, which is different from the affirmative approach of the Vedic Aryans. This perception of the world as suffering led the Hindu mind to escape the world of suffering through renunciation (Jhingram, 1989). The point is that both the orthodox and heterodox schools of philosophy have some similar ideas and at the same time diverged from each other by taking different approaches.

The third stage is represented by the Dharma-shastras, which include *Dharmasutras* and *Smritis* and *Puranas* as well as India's two great epics, *Ramayana* and *Mahabharata*. It took Hindu scholars from around the 5th century B.C. to the 15th century A.D. to write and finalize these texts. They are not well-defined systems of philosophy or religion but instead express very important traditions that have become coextensive with the entire body of Hinduism. These texts attempt to synthesize the Vedic and the philosophical tradition. Both the life-affirming standpoint of the Vedic tradition and the world-renouncing philosophical tradition prevail in Hinduism at present. The virtues of both the tradition and the philosophy are integrated, as evidenced by the advancement of the ideal of *ahimsa* (non-injury) in Hinduism (Jhingram, 1989).

The fourth stage of the religio-cultural tradition takes place with the *Bhakti* (devotional) movement. The *Bhakti* movement started with the Alwar saints of Karnataka in the last centuries before Christ, extending from there to Maharashtra in the west and then to the north. This movement influenced Hindu thought until the 16th century in the common era. According to Jhingram (1989), 'Its greatest contribution to Hindu morality lies in its rejection of the relevance of caste distinctions in determining the worth of a person' (p. 20). This theistic and devotional tradition follows the monistic or non-dualistic *Vedanta* philosophy and worldview, combined with the devotional approach to religion (Jhingram, 1989).

The fifth stage begins with the modern age in Indian history. After centuries of British rule, there was an unparalleled intellectual and spiritual awakening in Hindu society in the first half of the 20th century. Social reformers like Raja Rammohan Roy paved the way for a Hindu renaissance, followed by spiritual and moral leaders such as Swami Dayananda Saraswati, Ramkrishna Parahamsa, Swami Vivekananda, Sri Aurobindo, Ramana Maharshi, Ishwara Chandra Vidyasagar, and Mahatma Gandhi. Many more social reformers, intellectuals, political leaders, and scientists have contributed to moving India forward from the oppression of the previous several centuries (Jhingram, 1989). Interestingly, modern scholars of Hinduism have found fault with the social reformers, arguing that they continued to be influenced by Western ideas of colonialism. Nonetheless, there has been a positive transformation of Hindu ideals, models, and institutions in contemporary India that needs to be highlighted (see, for example, Richard Weiss).

This analysis of the different stages of Hindu religion and culture shows the dynamic nature of Hinduism as it has evolved along with societal and cultural changes. In the next section, I analyze the answers given by some scholars to the question of whether there is a Hindu ethics comparable to ethics in the West.

6.2 Comparison of Hindu *Dharma* and Ethics in the West

Scholars have raised the question whether there is an ethics in Hinduism similar to the way ethics is a branch of philosophy in the West. In her article, 'The Dharma of Ethics, the Ethics of Dharma: Quizzing the Ideals of Hinduism', Dhand (2002) argues that the Hindu ethics of *dharma* might have some similarities with ethics in the West, but it is also very different. Dhand focuses on the universalizability criterion found in the

West – in particular, Kant's view that gives value to the general rather than the particular – in order to compare and contrast it with *dharma* in Hinduism and decide whether it can be viewed as ethics in the Western sense. Dhand concludes that there are different levels to *dharma*, including one level that incorporates universalizability in terms of how to behave towards all human beings, but it also has another level pertaining to how to behave as a particular person in particular relationships.

Creel concludes in his 1977 work *Dharma in Hindu Ethics* that there is no body of literature in Hindu thought that corresponds to ethics in the West; he quotes other scholars like Raju, Devaraja, and Hopkins who hold a similar view. For Perrett, author of *Hindu Ethics: A Philosophical Study* (1998), ethics is concerned with two questions: What ought we to do and why ought we to do it? The theories of the Good in Hinduism have been highlighted in great detail by every generation of Indological scholarship. Dhand (2002) posits: 'It is the theories of the Right in Hinduism that concern us here' (pp. 349-350). Theories of the Right deal with the principles of behaviour and how people relate to each other and the world.

In Western moral philosophy, the Good is viewed in different ways: as happiness (by Aristotle), pleasure (by Mill), union with God (by Aquinas), or the good will (by Kant). Each of these philosophers developed their theories from these values. Although they differ in the method they use and the moral codes that follow from these values, there is general agreement among Western moral philosophers that all people whose reason is not impaired are moral agents. As Dhand notes, 'they share certain pervasively moral characteristics – humanity, moral sense, reason, and so forth' (2002, p. 350).

Dhand analyzes whether similar assumptions can be made about Hinduism. As Hinduism is diverse and has a number of authoritative works of literature covering different genres, it is difficult to find a definitive answer to these questions. She argues that the consensus of post-Vedic philosophy can be about the good: *moksha* (freedom, liberation, and release). Schools of philosophy like *Samkhya* and *Yoga* and *Vedanta* are in agreement about what is viewed as good. 'Classical Hindu thought, therefore, is primarily soteriological, focused on the telos of *moksha*, and philosophical arguments are generally rehearsed within the parameters of this religious frame' (Dhand, 2002, p. 350). Due to this intimate relation between religion and philosophy, many Western philosophers believe that Indian philosophy is not philosophy but rather religion.

Dhand addresses an oft-stated criticism of Indian philosophy that Hinduism does not recognize the universal principles of Right, as it is based on the hierarchy and division of *varna* or caste. By exploring various texts in

Hindu literature that are didactic – like the *Gita,* the *Ramayana,* and the *Yogasutras* – she explains the different levels of Hindu *dharma* or righteousness. She then shows that there is very little similarity between *dharma* and the discipline of ethics in Western philosophy. In Hindu literature, she argues, there is universal as well as particular morality, and it operates at different levels. Personhood is based on different factors like one's vocation or caste as well as other factors that might be hierarchical and relational. In Hinduism, differences among people are based on the 'idea that people are psychologically different' (Dhand, 2002, p. 352). The particularity of morality is built into Hindu classifications of action.

According to Dhand, *Dharma-shastra* literature (the treatises on *dharma*) are concerned with worldly issues, not soteriological issues. One might think that *Dharma-shastras* are concerned with ethics, but she says it is not that simple because in Hinduism, law, ethics, and soteriology are intertwined. Further, Hinduism 'acknowledges the levels of the Right' (p. 351). She notes two levels of ethical formulations: 'those directed at sophisticated intellects who are oriented toward the eternal, *moksha;* and those directed at the less-evolved majority, oriented towards the temporal, *samsara*' (351). In Hindu tradition, the agency of a person is dependent not only on one's mind but also on other factors such as one's position in society, including one's vocation or stage of life, one's age, and one's seniority in the various hierarchies of caste. Thus, codes of conduct cannot be generalized.

In Hindu tradition, the differences acknowledged among people range from physiological makeup to the paths chosen toward salvation. The *Samkhya* school of philosophy, for example, believes that nature or *prakriti* has three qualities or *gunas: sattva* (lightness), *rajas* (energy), and *tamas* (heaviness). People have different ratios of these three qualities, and that is what makes them unique. All the later schools of philosophy have incorporated this viewpoint. Thus, someone who has high energy is suitable for certain kinds of work, while someone who has a high degree of lightness is suitable for other types of work in society. And depending on how each individual is, the guidance of *dharma* is different too. In the soteriological strand of Hinduism, it is believed that ultimately, we are Selves or *Atman,* without any qualification. This is what true identity is. Most people become engrossed in the distracting world, living with the illusion that they are lawyers, teachers, doctors, etc. Those who recognize the Self are rewarded with salvation or liberation from the cycle of birth and rebirth by achieving oneness with the Absolute or *Brahman*. As Dhand (2002) encapsulates: 'The aim of religious practice is to discover our primordial and inviolable personhood, and to intuit its difference from the contingent identity that usually passes for personhood in the ordinary course of our lives' (pp. 353-354).

Different schools of Indian philosophy use other terminologies like *purusha* or *jivas*. Salvation or liberation can be achieved in different ways, and the *Gita* mentions three important ones: the way of knowledge (*jnana yoga*), the way of devotion (*bhakti yoga*) and the way of action (*karma yoga*). In Hinduism, the code of conduct that applies to everyone – irrespective of one's caste, class, or gender – can be considered ethical. Hindus 'stipulate the diligent care of other beings, scrupulous of the cardinal ethic of non-harmfulness first and foremost, as well as truthfulness, charity, patience, self-restraint, and compassion. They are geared toward self-cultivation' (Dhand, 2002, p. 355). Ethics and self-cultivation are the same thing, since caring for others cultivates the self.

6.2.1 The Gita and Dharma

In the *Gita,* the moral dilemma that Arjuna faces, as interpreted by Dhand, is not the choice between violence and nonviolence, as explained by some scholars, but between two sets of particular *dharma*: his *kuladharma* (*dharma* of the family) and *varnadharma* (*dharma* of the caste). The *dharma* of the family requires Arjuna not to harm his family. His *dharma* of the warrior or *kshatriya* caste requires him to kill anyone who is an enemy involved in injustice. Which *dharma* is more compelling? For ordinary unenlightened people, the particularities of the situation become very important. For enlightened people, it is one's duty to eradicate the injustice or evil to bring order. There is also common *dharma* which applies to everyone in all places, irrespective of gender, caste, or stage of life. This is called *swadharma, samanyadharma,* or common *dharma*. It outlines duties that must be followed by everyone at all times. The codes of common *dharma* are stated in classical texts. For example, the epic *Mahabharata*, as Dhand notes, includes passages about the restraint of anger, truthfulness of speech, agreeable nature, forgiveness, begetting children with one's own wives, purity of conduct, avoidance of quarrel, simplicity, and the maintenance of dependents.

The *Bhagavad Gita*, which is a part of the *Mahabharata,* has a significant impact on the general Hindu population of India due to its moral messages. The two outstanding features of the *Gita* are its possibility of following both paths of liberation – the pre-Vedic asceticism and the performance of moral and social duties. The *Gita* arrives at the reconciliation of these two traditions, which are in competition with each other. The way it is resolved in the *Gita* is through *nishkama karma*, or doing the action without thinking about the consequences. It proposes that doing the action as a duty is what is essential. The *Gita* says in the second teaching, verse 47:

> Be intent on action,
> not on the fruits of action;
> avoid attraction to the fruits
> and attachment to inaction!

One of the most important ethical questions raised in the *Gita* is whether Arjuna should kill his cousins and relatives for the sake of his rightful sovereignty. In the end, Krishna advises Arjuna to do his duty as a warrior without thinking about the consequences, as the evil caused by the Kauravas, has pervaded the world, and it is essential to eradicate the evil of injustice at any cost. Kauravas wanted to control the kingdom, which was not rightfully theirs. The kingdom belonged to their cousins, the Pandavas. Arjuna, the leader of the Pandavas, was overcome with grief seeing all his cousins and other relatives in the battlefield and reflecting whether it was worthwhile to kill his relatives for the sake of their own (Pandava's) kingdom.

At first glance, following one's duty for the sake of duty sounds like Kantian ethics. But it should be pointed out that the rational-universalizable formulation of Kant is lacking in the case of the *Gita*. Mohanty (2007) remarks that while for Kant, the principle of universalizability is the form of the test, in the *Gita,* Krishna does not use this test of universalizability. In the *Gita*, 'it would appear the form is simply prefixed to the already available content' (p. 70) of *dharma* which already states that the *kshatriya* or warrior caste is going to follow his caste duty to fight a battle if the cause is righteous. In the *Gita,* Krishna adds that in order to achieve *moksha* or liberation, this duty should be followed with non-attachment. One has to do their duty according to their disposition and caste.

Mohanty goes on to state that in the *Gita,* the spiritually perfected individual moves above the three *gunas* of *sattva* (wisdom), *rajas* (activity), and *tamas* (darkness). A virtuous man who follows *dharma* is under the influence of *sattva.* Krishna accepts the metaphysical theory of the *Samkhya* school of philosophy and argues that, although under the influence of *sattva*, one is still within *prakriti* which binds itself to the self.

> The theory holds however that there is a core of my being which I reach only when I strip away all the products of nature in me, my body, the

> sense organs, the inner sense or the mind, the ego-sense, the intellect which it calls *Purusha*, whose nature is pure consciousness, and which is an uninvolved spectator. (Mohanty, 2007, p. 71)

It is only through this process that the individual can transcend *dharma*. Thus, the higher goal is sought out due to the laws of scriptures, not reason, as the Kantian view would suggest. Even though the general population cannot transcend *dharma,* it provides the social ethic that the society should live by.

6.2.2 The Ramayana and Dharma

The epic *Ramayana* is the story of Rama, the eldest of four sons of King Dasaratha of Kosala. King Dasaratha was generous and righteous and served his subjects well. But he made a misguided promise to a junior wife (of three) to disinherit Rama and banished him from his kingdom for fourteen years. At first, King Dasaratha hesitated to banish his most deserving son, but Rama followed his father's order, though his mother and brothers did not want Rama to be in exile. Rama's wife Sita and his brother Lakshmana accompanied him for the fourteen years; they faced many misfortunes, one of which was Sita being kidnapped by the demonic King Ravana and being held captive for one year. Rama fought a ferocious war with the demons before he could bring Sita back. After that long struggle, at the end of fourteen years, Rama returned to his father's kingdom, and there was much celebration for the completion of his exile. But then, Sita's womanhood was questioned due to her being kidnapped. Rama 'yields to what he considers his *dharma* as a king beholden to his people by sending Sita to the forest, where she bears twin boys' (Dhand, 2002, p. 361).

Dhand says that in the *Ramayana,* some kind of common *dharma* has been developed, but it does not stop being relational. It is from the relational that the common *dharma* or ethics is derived. 'Perhaps paradoxically, one is taught how to behave toward all human beings by learning how to behave as a particular person in particular relationships' (2002, p. 360). After meticulous analysis, Dhand concludes that Rama's ideal character in the story of *Ramayana* has been guided by the common *dharma* and at the same time by relational *dharma*: 'It would appear that if there is one ethic informing all of Rama's actions in the *Ramayana*, it is a stoic courage, a willingness to bear all hardship for an uncompromising commitment to *dharma*' (p. 365).

6.2.3 The Yogasutra and Dharma

Another classic text is Patanjali's *Yogasutra,* which prescribes eight steps to achieve self-realization. Of the eight steps, the first two – *yama* (restraint) and *niyama* (discipline) – lead to moral perfection; only when moral perfection is accomplished can one move to the third and further on to the eighth step, in hierarchical order. Dhand (2002) explains that *yama* involves paying attention to the five ethical principles of *ahimsa* (non-injury), *satya* (truthfulness), *asteya* (non-stealing), *brahmacharya* (chastity), and *aparigraha* (non-possessiveness). The second step is *niyamas,* which involve habits that must be cultivated. The five precepts to follow are *saucha* (cleanliness), *santosha* (contentment), *tapas* (stoic endurance of discomfort without complaint), *swadhyaya* (habit of study or reflection), and *Ishwarapranidhana* (meditation of *Ishwara* or God). These are common moral codes that apply to anyone irrespective of their class, caste, gender, etc.[4]

From this analysis of *dharma* in Hinduism, it is apparent that Hindu ethics is very different from ethics in the Western world. Hindu ethics of *dharma* as social duty, with its many implications, is the backbone of Indian culture. The most popular religious scriptures such as the *Gita* and the two epics, the *Ramayana* and the *Mahabharata* – which are known, read, and performed in countless movies – have a major impact on day-to-day life in India, though sometimes in an ambiguous way, particularly when the ideals of *dharma* are not mirrored in actions. This is where the disconnect arises between the teachings of the religious scriptures and the way people actually act. In our time, people need to be educated about following the rituals with an awareness of the broader issue of sustainability.

6.3 Hindu *Dharma,* Ecology, and Sustainability

An interesting question to ask is whether Hinduism as a religion, philosophy, and ethics has any role to play with regard to ecology and sustainable development. In the last three decades, scholars have written extensively about the connection between religion and ecology, pointing out various aspects of Hinduism that have influenced an ecological viewpoint. The next

4 These eight steps are hierarchical. The third to eighth steps are *asana* (posture), *pranayama* (breath control), *pratyahara* (withdrawal of the senses), *dharana* (concentration), *dhyana* (meditation), and *samadhi* (pure contemplation).

question one might explore is: what is the connection between *dharma* and sustainability? Scholars have debated this question's finer points.

Yamini Narayanan argues that experts in the fields of science and technology must embrace the value of life in order to bring about a sustainable viewpoint that should be followed by everyone. Narayanan points out that some scholars, such as Agarwal (2000), claim that the practices of Hinduism itself contribute to environmental degradation in India, as Hinduism 'looks into the self, emphasizing the *atman* as the key to spiritual ascent' (as cited in Y. Narayanan, 2013, p. 302). Pollution in India is viewed in an individualistic way, in relation to the self, and has not created 'a culture in which institutions could grow' (as cited in Y. Narayanan, 2013, p. 302). Whereas Agarwal claims that the interconnectedness of the self and the environment is weak to begin with and is worsened by our contemporary materialistic society, Narayanan challenges this notion and argues that it is all the more important now to revive a green sustainable self in Hinduism to solve our current problems of environmental destruction.

By analyzing the concept of self as summarized in Hinduism's *purusarthas* or four aims of life – *kama* (sensual desires), *artha* (accumulation of wealth), *dharma* (righteousness and social morality), and *moksha* (liberation) – Narayanan shows how these four aims lead to a higher or greater self. The four aims can be seen as four stages of life. Beginning with puberty and the basic human nature to want material things, we progressively move to the stage of liberation (or self-actualization or transcendence) by following moral precepts not only at the individual level but also the collective level, thereby leading to a higher or greater self. She writes: 'I suggest rather that self-conscious experience of the good life is a vital way to live the meaningful life and thus, ultimately, ecological and social sustainability' (2013, p. 303). It is important to note that the *purusarthas* in Indian philosophy can be applied within theistic, atheistic, and agnostic frameworks.

The third stage or aim of life, *dharma*, has a spiritual as well as a secular dimension. It entails a code of conduct that should be practiced.

> Here the science and methods of sustainability may be enormously useful in informing the secular notion of *dharma*, while the metaphysical aspect of *dharma* may assist in enabling sustainability. I thus propose that the third stage of *dharma* is where the most profitable connections between Hindu religion and sustainability may be made. (Y. Narayanan, 2013, p. 304)

Narayanan remarks that in India, religion and politics are intimately related and therefore, the political journey can lead to the goal of self-realization.

The World Trade Organization (WTO), the International Monetary Fund (IMF), and World Bank follow a neo-liberal capitalistic viewpoint and thereby fail to take into account the moral aspect of life. The first two stages of *kama* and *artha* do not imply that they should be pursued without any restrictions. Actually, it is *dharma* (or the moral code) that keeps the sensual desires and wealth in check, by its moral injunctions in preparation for the fourth stage of *moksha* (liberation or transcendence). Having wealth is encouraged, but it is also morally wrong to deprive others of their wealth, as stated in the *Mahabharata* (12.8.11). *Kama* has to do with sensual desires, but it also includes enjoyment appropriate to the objects of the five senses – hearing, feeling, seeing, tasting, and smelling – controlled by the mind that is driven towards the self. Thus, the aims of *artha* and *kama* do not justify the exploitation of economic, social, and environmental assets. The *Bhagavad Gita* claims that it is by obeying the first three stages with dedication that one can achieve the fourth stage of *moksha*, which releases the self from the pain and suffering of this world and thereby sets it free from the cycle of birth and rebirth (Radhakrishnan, 2005; Y. Narayanan, 2013).

Narayanan also notes that *purusarthas* echo with enlightenment (Y. Narayanan, 2013; Kuppuswamy, 1977) and Clive Hamilton's notion of 'inner freedom, the freedom to act according to one's own considered will' (2008, p. 218). If enlightenment means a state of complete self-awareness, then enlightenment can be viewed as a state of 'inner freedom.... guided by the moral self' (Hamilton quoted in Y. Narayanan, 2013, p. 309) The self must be guided by morality when pursuing wants and desires.

Artha (or the pursuit of wealth) includes not only one's desire of personal wealth but also the political duty of pursuing commerce, which is an essential task of a householder. In this sense, *artha* is secular. Thus a householder, in the second stage of life, must be involved with a healthy society in which people pursue economic growth and political participation. Wealth must be spent in an enlightened way so that it leads to mature self-understanding.

Development scholars have written extensively on how economic growth and economic justice as well as sustainable economic development must be integrated with ethics-based values. According to former World Bank economist Alfredo Steir-Younis (2001), because economic values and economic decisions are the only things that matter in this global society, this leads to dysfunction at all levels. He explains that a 'healing experience will come about when societies are able to reconcile economics with spirituality' (Steir-Younis quoted in Y. Narayanan, 2013, p. 316).

Kama (or desire) is generally understood as sexual desire, as stated in the *Kama-shastra.* But there are other non-sexual aspects of desire considered in Hinduism that give desire a spiritual aspect rather than a purely secular aspect. With regard to non-sexual desires, Narayanan observes that in the current era, economic growth with material wealth is promoted in a capitalistic system so that poverty is going to be reduced and eventually eradicated. She notes that 'intense focus on accumulation of material and physical luxuries has led to what can be described almost as psychological affliction: affluenza' (Y. Narayanan, 2013, p. 317).

Once a person's well-being is achieved upon reaching a certain standard of living, they will turn to non-material pursuits like community and the natural environment. It is clear that within Western countries and among middle-class people in developing countries, this is not happening. Hamilton and Denniss have shown that to solve the problem of poverty, the problem of affluence must be resolved first, because poverty and environmental destruction are integrally associated with over-affluence and overconsumption. This is where Hinduism points out that *kama* has to be channeled to the dharmic expression of desire that entails self-reflection. Desire leads to bondage and the cycle of birth and rebirth, hindering the achievement of *moksha.* Inner freedom is possible only through morality. As Hamilton says, 'being free and being moral are inseparable' (as cited in Y. Narayanan, 2013, p. 319).

Moksha, according to Narayanan, means inner freedom. She argues that *moksha* is a 'self-aware sustainable self that in turn makes for a sustainable society' (2013, p. 309). In one sense, the self might be considered as a small component of sustainability. But in another way, it is at the core of sustainability: 'If the ecological crisis may be viewed as fundamentally a spiritual crisis, then self-development of the individual human self becomes the crux of global sustainability' (2013, p. 309).

In the classical texts of Hinduism, *moksha* is described as liberation, meaning when liberation is achieved, one becomes free from the bondage of birth and rebirth. According to Narayanan, 'Moksha essentially means the release of the human consciousness from bondage into enlightenment' (2013, p. 309). It is the exclusive attachment to material things that causes misery and meaninglessness, and it is the transcendence of the material realm to the metaphysical realm that leads to inner freedom. From the *Brihadaranyaka Upanishad:*

> Lead me from non-being to being
> Lead me from darkness to light
> Lead me from death to immortality (3:28)

Immortality is the recognition of the interconnectedness of all things. It is the consciousness of being one with the cosmos (Chaudhuri, 1979). Narayanan claims that in this sense, *moksha* can be achieved in one's lifetime, not after death as stated in some classical texts. *Dharma* 'is where the most productive union between Hindu metaphysics and sustainability may be made' (Y. Narayanan, 2013, pp. 311-312). *Dharma* is translated in many different ways: as duty, law, righteousness, justice, ethics, etc. As *dharma* prescribes a moral code of conduct in response to one's desire and behaviour, it allows for the resolution of conflicting wants and needs, desire and duty. 'This code may be usefully informed by both metaphysical injunctions as well as the voluminous body of knowledge that is now available on various aspects of economic, ecological and social sustainability' (Y. Narayanan, 2013, p. 312).

The word *dharma* was used for the first time in *Rigveda*, one of the four sacred Vedic texts. Nadkarni (2011) points out that there are different meanings of *dharma,* 'in different contexts – as synonym for satyam, duties, and ritual obligations' (p. 219). The word *dharma* literally means to uphold or sustain, and it is used in a broad sense in Hindu philosophy. In *Atharvaveda,* a collection of ancient spells, prayers, chants, and hymns, it is said that the world is sustained by *dharma*. In Jainism and Buddhism, the word *dharma* is used in an ethical sense, not in a ritualistic way. *Dharma* has different meanings, yet this word is used in all Indian religions in an ethical way. Narayanan points out that sustainable development can be related to the *Bhagavad Gita's* commentary on *dharma*. The *Gita* proposed that the ultimate reality or *Atman* can be attained through *jnana yoga* (the way of knowledge) and *karma yoga* (the way of action). The way of knowledge is a way of giving up one's material desires in favour of the way of knowledge of the ultimate reality. *Karma yoga* leads to the same end by acting according to the moral code. Most people generally follow the path of action, which Narayanan explains relates to the path of material and sensuous experience. The lesson of the *Gita* is that the concerns of the body are not as important as the concerns of the spirit or *Atman*. It is with the primary focus on the spirit that detachment – doing one's action without thinking about the consequences – ensues, with the guidance of *dharma*. Krishna advises Arjuna that the true nature of the self is different from action. He does not advise Arjuna to give up action but rather to act according to the moral law prescribed by *dharma*. It is the guidance of *dharma* that does not take people to the extreme paths of despair or indulgence in vanity. In order to achieve *Atman* or the true self, Arjuna must act in a desireless way, without the attitude of possession, as it is desire that makes people possessive of their object of desire (Y. Narayanan, 2013).

As noted, it is in the third stage of life – *dharma* – that a connection can be made between sustainable development and Hinduism as well as between science and spirituality. Narayanan points out that 'The challenge is to be conscious that desire and consumption are informed by *dharmic* sustainability, so that an important route to self-realization is not compromised' (2013, p. 313). She suggests that the experts on sustainability should combine their scientific and technological expertise with *dharma*, which provides the spiritual dimension in order to reform the social and political aspects of sustainability. Narayanan concludes her article by saying that 'for human beings to be happy and liberated, for *dharma* to be truly lived, and sustainability to become a conceivable reality, development policy will have to think beyond technical strategies and analyses and begin to consider, in philosophical and pragmatic terms, the meaning of life' (2013, p. 320).

This analysis of Narayanan's article shows that the Hindu concept of *dharma* might have a transformative power in the context of sustainable development in light of the ecological crisis that not only the people of India but the entire world face. It calls for ethical values to be linked to the scientific, technical, economic, and political levels of society by responding to biocide (the destruction of living organisms) and geocide (the destruction of the planet), two terms that were coined by Thomas Berry.

Narayanan's claim can be viewed as providing an important dimension to the sustainable development (SD) policy debate in India by emphasizing the aspect of the meaning of life in the existing scientific, economic, and political debates. The three aspects of the United Nations' sustainable development goals (SDGs) are economic, social, and environmental. Public policy debates about the social aspect of sustainability development must shift its priorities from economic growth to not only material well-being, by providing opportunities for all people, but also to mental health, by emphasizing the moral codes of *dharma* based on the Hindu tradition, which is well-known to the large population of India.

This reiterates the point made in Chapter 5 about the core values of development ethics, in which the ethical component of 'well-being' is considered the overarching foundation. Here, in the context of India, the ethical component can be taken a step further with the Hindu ethics of *dharma*. As has been pointed out in this chapter, the ethics of *dharma* is different from the Western sense of ethics, but they overlap in terms of the common codes or universal code (in the Western context). In addition, Hinduism has individual codes that have to be followed depending on one's position in society, caste duty, the stage of life one is in, and so on.

6.3.1 Hindu *Dharma* and Applications in Ecologically Sustainable Development

The question that arises now, as posed by another scholar, Vasudha Narayanan, is: 'What, if anything, can Hindu tradition say about this looming environmental crisis? Are there any resources in the Hindu religious and cultural traditions that can inspire and motivate Hindus to take action'? (2001, p. 179). In her article, 'Water, Wood, and Wisdom: Ecological Perspectives from the Hindu Tradition', Narayanan highlights another important religious perspective: how the Hindu traditions translate into action. After documenting many classical texts of Hinduism on the harmony of humans and nature as well as another set of texts based on *dharma* and ethics, she points out that people have taken to such texts selectively.

In India, there is a deep relationship between social structures and behavioural patterns. The general population loves the ancient Hindu texts like the *Puranas,* which are about the numerous Hindu deities, and the epics *Ramayana* and *Mahabharata* – many of which are shown in plays and movies. In many Hindu texts, plants and trees are given significant value. There is also a close correlation between *dharma* and the ravaging of the earth. When *dharma* declines, human beings begin to despoil nature. The Hindu texts that focus on *dharma* are complex; they do not advise people to be passive and accept the life-negating philosophy of the end of the world. but also promote improvement in the quality of life so that happiness can be achieved in this life. In many temples, the blessing uttered focus on 'human happiness in this life, on this earth' (V. Narayanan, 2001, p. 181). And though this is the case, Hindus still pollute the environment.

So how has ecology been addressed in Hindu texts? To explore the answer to this question, it is important to note that there are many Hindu traditions and that there is no one authoritative book on Hinduism, as stated earlier. Another point is that the *Puranas* and the epics are more influential in Hindu culture, but the philosophical works such as the *Upanishads*, which came out of the oral tradition, are not known to the general population. The texts on right behaviour like the *Dharma-shastras* are selectively followed, and popular customs are given the same weight as religious laws. All these texts have promoted the value of *dharma* and *bhakti* (devotion). At the popular level, *dharma* is important in the sense that the codes were conveyed through the epics, which always communicated a moral message of acceptable behaviour. Later, during the period of British colonization, the people of India relied more on texts of the law. Over time, due to the influence of mass media, texts became more important than the oral tradition and the customs of the community. Today, many Hindu texts can be considered as resources for environmental philosophy (V. Narayanan, 2001).

As Narayanan points out, there are many religious resources found in the Hindu tradition that can be used to raise awareness and that urge action with regard to the environmental crisis in India. Many ethical texts emphasize the importance of nonviolence towards all creatures. The highest form of righteousness or *dharma* is nonviolence in thought, speech, and action. One can argue that normative nonviolence promotes biodiversity, provided that people follow it. Similarly, Manu, the lawgiver, proclaims that impure objects should not be thrown into the water. The protection of sacred groves, the sacrality of land, and India personified as Mother Earth have all been important ideas in political thinking. In songs, as Narayanan points out, 'India is personified and extolled as a compassionate mother-goddess filled with forests, filled with sanctity that should not be violated' (2001, pp. 183-184).

In Hindu philosophy as it has evolved from the *Upanishads,* nature is viewed in different ways. Nature is sacred, and in some schools of philosophy, *prakriti* (nature) is divine immanence and has potential power. These links have been explored in the quest for indigenous paths to solving the environmental crisis. In the *Bhagavad Gita,* we see the vision of the universe as the body of Krishna, an incarnation of God Vishnu. One interpretation of this is to convince Arjuna of the supremacy of God. The philosopher Ramanuja and others have understood passages from the *Upanishads* as depicting the correct relationship between the Supreme Being and creation, and they emphasize the immanence and transcendence of the Supreme Being. According to Ramanuja, the universe, composed of both sentient matter (*chit*) and non-sentient matter (*achit*), pervades a non-sentient body (*achit*) – and in the same manner, Vishnu pervades all souls, the material universe, and time. The name Vishnu means 'all pervasive' (V. Narayanan, 2001, p. 185). Vishnu is the name that is given to the Supreme Being or *Brahman.* Ramanuja, the founder of the Indian school of philosophy, *Vishishtadvaita* or qualified nondualism, asserts that diversity results from underlying unity. Reality of all creation and its divinity are based on scriptural passages. The created world is as real as Brahman. The reality of all of creation fluctuates with divinity. This vision of an organic connection between the Supreme Being and all created beings asks us to view the world with wonder and respect. 'If the entire universe is divine, how can we bring ourselves to pollute it? Ramanuja's is only one of many philosophical visions of the universe that has bearing on the ecological enterprise', as Narayanan indicates. (2001, pp. 186-187)

6.4 Ways Hindus Connect to Nature

In addition to the selectivity with which Hindus adhere to their sacred texts, Vasudha Narayanan points out that there are a variety of ways that Hinduism is practiced in India as well as outside of India. In India, temples play a major role in people's day-to-day lives. Many temples are able to exist due to the endowment of their devotees. In addition, there are many gurus revered by millions of Hindus due to their supernatural qualities and the organizations funded by their devotees. Narayanan describes the example of the Tirumala-Tirupati temple in Andhra Pradesh, South India, which greets visitors with this statement on their billboard: 'Trees when protected, protect us', in reference to their reforestation work. The statement is taken from the Laws of Manu, which say that when protected, *dharma* or righteousness protects us. Narayanan also mentions the ecological work taking place cleaning up the pollution in the Ganges and Yamuna rivers, highlighting the example of Sathya Sai Baba, though deceased in 2011, one of the most influential gurus in India to this day. In 1994, he announced that his Sathya Sai Central Trust would undertake a water supply project. Sai Baba emphasizes the connection between rivers, religion, and morality. In Sai Baba's view, a lack of water leads to lack of morality. As Narayanan reminds us, while the ecological impact of Sai Baba's actions might be debated, as a teacher he has great power in influencing his devotees to change their behaviour.

The Bishnoi community of the state of Rajasthan in the northwestern part of India recognizes the eco-religion of environmental conservation by protecting trees and animals as their religious obligation (Sharma, 1999). Their basic philosophy is the idea that all living beings have the right to survive. They have a set of 29 rules, with eight tenets that prescribe biodiversity and good animal husbandry (see 29 rules in https://en.wikipedia.org/wiki/Bishnoi). The killing of animals and trees is prohibited by Bishnois. In their mission to protect all life forms, they do not use firewood that have insects in them. They collect rainwater in underground tanks in their desert area, as water is precious. Bishnoi communities also exist in the states of Haryana and Punjab. Their environmental awareness and commitment to protecting all living beings is notable compared to other religious communities in India (Prasad, 2013).

In his book *Dharma and Ecology of Hindu Communities: Sustenance and Sustainability*, Jain (2011) notes that the concept of *dharma* has evolved from its original meaning not only semantically but also in its linguistic expression of religiosity. *Dharma* is used in Indian communities in their daily life as virtue, ethics, duty, and ecology. According to Jain, 'It can help spread environmentalism because of multivalent significances of dharma in the Indian context' (2011, p. 105). The word *dharma* is translated as religion, but in Hinduism it is used to mean virtue, righteousness, and cosmic law and order. Following Dhand's argument as stated in the previous section, Jain suggests that *dharma* provides a universal ethical perspective, and this ethical interpretation can be viewed as the foundation for activism in general and for ecology in particular. For rural people in India, there is no significant difference between religious ethos and the ecological order: both are described as *dharma*. Jain analyzes the Bishnoi, Swadhyayee, and Bhil communities by showing how the lives of people in rural societies should be interpreted by using *dharma* rather than the Western conceptual framework of religion. Indian society, like any other, desires order; *dharma* describes this order and is the organizing principle of the Indian belief system. Jain points out, 'Thus, morality and natural phenomena are connected and interdependent' (2011, p. 113). In the human world, *dharma* represents social duty; in the natural world, it represents the intrinsic properties of natural entities; and in the supernatural world, it relates to religion.

Pandurang Shastri Athavale, the founder of the Swadhyaya movement, stressed that actions done with devotional motives can lead to the achievement of *moksha*. He believed that dharmic duties are not opposed to *moksha*. The world is divine, as God resides in the world (Jain, 2011). Athavale led a householder's life, and his followers are also householders. This community developed in the western states of Maharashtra and Gujarat when India began to modernize and industrialize. Athavale combined the use of technology with *dharma*. Both Athavale and Jambheshwar, the founder of the Bishnoi sect, developed their own ethical framework for their communities and validated them within the broader framework of *dharma* by supporting and quoting the epics, the *Ramayana* and the *Mahabharata*. Swadhyayees and Bishnois are vegetarians and farmers, and they show devotion to water, earth, trees, and cattle (Jain, 2011). Their communities are not socio-politically or monolithically organized.

Guru Jambheshwar preached *karma yoga* (the way of action). Like the Swadhyayees, Jambheshwar did not consider modernity an obstacle. He led a life of the celibate ascetic and performed spiritual acts in the desert of Western Rajasthan, but his followers were and are householders. Many Bishnois have held key positions in local government handling the department of forestry and animal protection (Fisher, 1997; Jain, 2011). By following the 29 rules of the Bishnois, as stated above, the Hindu ritual of revering animals and trees is preserved. In the Bishnoi tradition, idol worship is rejected, though they consider themselves 'a caste-group within the larger Hindu community' (Jain, 2010, p. 128).

The Bhils are hunter-gatherers and meat-eaters residing in Rajasthan, Gujarat, and Madhya Pradesh – the western Deccan regions and Central India. They are still seen as a tribal people on the fringes of mainstream society (Jain, 2011). This community does not have a charismatic founder or any strict guidelines to follow. They continue to maintain their own sacred groves based on their faith in the deity associated with a particular grove.

After studying these three communities in detail, Jain concludes that the 'Dharma Ecology of these communities offers a unique avenue for approaching environmental restoration today' (2011, p. 4).

Returning to Vasudha Narayanan's article, she has demonstrated that the classical Hindu texts and the cultural ways that Indians have taken up selectively to follow in their everyday lives can be 'the greatest potential resource for ecological activists in India' (2001, p. 202). She makes a plea for a 'shift in our perspective from *tattva/moksha* texts to the resources that have a more direct relevance to worldly behaviour' (2001, p. 202). She mentions the Dharma texts and devotional exercises in particular, noting that devotion to God Krishna or Mother Ganga has inspired some people to take action to supply safe drinking water, to clean up rivers, and so on.

The experts in the field of religious environmentalism whose views have been explored in this section show the different ways in which *dharma* has been approached by Hindus in religious, spiritual, and cultural ways in people's day-to-day lives – in most cases not due to environmental awareness but as a ritual and a religious view to follow. This demonstrates that relating to things that have direct relevance to their behaviour in this world is an avenue that can be taken to make Indians more environmentally aware. Instead of using the religious works in a selective way, people will then opt to protect the environment in a broader way, as protecting the environment and saving the planet's biodiversity can only help them in the end.

6.5 Influence of Symbolic Traditions on Some Environmental Cases

In answering the question: 'How have modern day Hindus drawn on the mythological and symbolic traditions of their faith to address our current ecological crisis'?, Kent (2016) considers three different strands of Hinduism:

1. The Brahmanical tradition, or 'great tradition' that was passed on by priests, scholars, monastic ascetics and gurus;

2. The popular 'little tradition' of worshipping nature, animal sacrifice, etc.; and
3. Neo-Hinduism, or reform Hinduism and renaissance Hinduism that emerged from reform movements to purify the 'backward' practices, such as widow burning and animal sacrifice, and make the Brahmanical tradition more accessible to the general population. (Kent, 2016, p. 291).

The Sramana tradition's *Advaita Vedanta* asserts the unity or non-duality of all existence. It advocates self-realization of the illusoriness of things within the universe. The innermost spiritual essence, the *atman,* is one with the ultimate reality of *Brahman.* As Kent points out, this worldview seems to have a positive relation to the environment, as it believes in the unity of nature. But Nelson (1998) argues in his influential essay, 'The Dualism of Non-Dualism', that *Advaita Vedanta* supports a stark metaphysical dualism in that it sees the *Atman* or spirit as integrated in the matrix of the material world – *maya* or *prakriti* – which is illusory in the sense of being ultimately unreal. In asserting that the only true and ultimately real dimension of reality is pure spirit or *Brahman, Advaita Vedanta* validates extreme asceticism, urging seekers to see the natural, material world as of 'no more worth than the excrement of a crow' (Nelson, 1998, p. 81). Nelson argues that *Advaita Vedanta* thus fosters an attitude of indifference to environmental decay and destruction by viewing the natural world as ultimately valueless relative to the absolute spirit (Kent, 2016).

Kent asks if this dualistic viewpoint is responsible for the pollution of the river Ganges. There is a contradiction between the way in which sewage, garbage, and industrial pollutants flow freely into the river and the religious and mythological belief of Hindus that Mother Ganges purifies all sins. Hindus believe in immersing the ashes of a deceased relative in the Ganges river, with the belief that the sins of the deceased will be washed away, 'adding both to its reputation as a site for the outpouring of divine grace and its pollution load' (Kent, 2016, p. 294). Kent highlights the Sankat Mochan Foundation, an environmental NGO actively involved in influencing city officials to keep the Ganges river clean.

Similarly, Kent presents research done by Haberman (2006), which shows that the Yamuna river is viewed as Mother Goddess, following from the *Bhagavad Gita* tradition that identifies Krishna as the God from whose body the entire universe has emerged. So the Ganges is associated with death, as the tradition of immersing the ashes of the deceased is followed, and the Yamuna is viewed as the goddess of love, associated with Krishna, as his lover and partner in creation. Both rivers are highly polluted and yet revered as Mother Goddesses (Haberman, 2006; Kent, 2016). Haberman concludes that the Bhagavata tradition, one of 18 great histories in the *Puranas*, can be a motivator for environmentalists to act, like the environmental movement in Vrindavan, the town where Lord Krishna grew up. It is regrettable that those who pollute the Yamuna river do not view it from an environmental perspective. India's transformation from a socialist economy to a capitalist one since the 1990s has worsened environmental destruction. Today, India is confronted with the massive pollution of its rivers, soil, air, and habitat as well as groundwater depletion, deforestation, climate change issues, among others.

It is interesting to note here that James (2014) argues, in response to Nelson's claim, that while it is true that the legendary philosopher Shankara or his followers of *Advaita Vedanta* did not write about the intrinsic worth of nature, it does not follow that this school of thought did not have any environmental concern. For *Advaita, moksha* is the ultimate reality, and the goals of economic well-being are dependent on duty or *dharma* that has to do with the visible world. Secondly, although Shankara holds a negative view of the visible world, the goal is to oppose attachment to the material world, as it is this attachment, due to ignorance, which leads to the cycle of birth and rebirth. Thirdly, by renouncing the natural world, followers of Shankara are neither exploiting earth's resources nor protecting them. Finally, James emphasizes:

> It is precisely by the renunciation of worldly ambition that some leaders of environmental movements have achieved the moral authority to address abuses of power that profit from the exploitation and destruction of environments upon which the powerless and disenfranchised depend. (2014, p. 14)

Another philosophical viewpoint that James states is Ramanuja's 'qualified non-dualism' or *Vishishtadvaita*. This school, like Shankara's *Advaita Vedanta*, believes that *Brahman* is the ultimate reality but that it is not distinct from the empirical world, which is a part of *Brahman*. *Brahman* is the unity of the differences that constitute the experienced world and the ground. Individual selves are modes of *Brahman*. 'All is *Brahman*' means that the things and the selves are the complete body of God, real qualifications of *Brahman*. The individual souls permeate the physical body, and God permeates all souls. For Ramanuja, devotion to God is an admissible means to acknowledge the presence of God in the natural world.

6.6 Is Hinduism Eco-Friendly?

The question of whether Hinduism is eco-friendly is debated by many scholars, as the harmony of humanity and nature has been a common theme in many Hindu religious scriptures from the *Vedas* to the *Upanishads* and in the two epics, the *Ramayana* and the *Mahabharata,* as well as in many Indian schools of philosophy. The perspective taken by Tomalin is that some of the readings of Hinduism support environmental concerns. But, she says, 'The contention that religious traditions are "environmentally friendly" is an *interpretation of tradition* rather than a *traditional interpretation*' (2002, p. 15).

Tomalin makes references to the worship of trees, groves, and rivers as found in Hindu scriptures, making the point that although nature is worshipped in these ways, it does not promote the worship of nature as a whole but only to some specific trees like bilva (which is sacred to God Shiva) and the aswattha (sacred to God Vishnu) as well as groves such as kavas. These are protected and worshipped due to a deity to whom the grove was dedicated; likewise, the Ganges is worshipped because it is believed that the sacredness will take away all impurity, even of pollution, as Alley (1998) discovered in her field work. According to Tomalin,

> A religious approach to the environment is a contemporary phenomenon which evolved alongside the increasing capacity of humanity, particularly since the industrial revolution, to manipulate the natural world on a massive scale with global consequences. By contrast, nature religion is the worship of elements of the natural world, most often with no basis in the ideas and values of contemporary environmental thinking. (Tomalin, 2002, p. 18)

Tomalin takes the term 'empty belly environmentalism', coined by Guha and Martinez-Alier (1997), to mean that in India, environmental movements are based on the survival and livelihood of the poor people. Such movements have two important characteristics according to Tomalin. First, when poor and rural people struggle over resources, the issue is viewed as one of survival, local autonomy, and cultural identity and not as an environmental issue. In her words, 'The concept of "environment" that we have in the West is often fairly meaningless in India' (Tomalin, 2002, p. 21). Second, there is a conflict of interest within India between people who have access to the political framework and are not directly affected by the situation, who try to impose their ideas on the local people without proper understanding, and the experience of local people in a situation.

She points out that this is prevalent not only in India but also in the West and globally.

Tomalin's first main point is that India's concern over the environment is not founded on the idea of religious environmentalism, as in the West. She points out that there are references to nature in the literature of Hinduism, but from that, it does not follow that it is necessarily religious environmentalism.

This discourse also advances the values and lifestyles of non-industrialized communities – in particular, the tribal or peasant cultures – as environmentalist. Tomalin mentions Calicott's point that the Chipko movement drew on Gandhi's principles of political action and the ideas of Hindu philosophy by organizing Indian villagers to protest against industrial deforestation (Calicott, 1997; Tomalin, 2002).

Tomalin argues that the Chipko movement was not grounded on any ideology. As she suggests, 'values and ideologies are being imposed upon the movement *post hoc* and at the level of popular participation the Gandhian label is inappropriate' (Tomalin, 2002, p. 14). She cites Guha who has pointed out that the Chipko movement has used religion for pragmatic reasons and not as a philosophical foundation. According to Tomalin, 'Whilst nature religion is a feature of Hinduism, it would be incorrect to call Hinduism *per se* a nature religion or "environmentally friendly"' (2002, p. 16). It can be argued that, to some extent, it is true that the women who followed the Chipko movement were primarily concerned with their right to livelihood which, as described in Chapter 2, gave them the strength to unite and fight for their basic rights. But it is also important to remember that the two leaders of this movement, Sunderlal Bahuguna and Chandi Prasad Bhatt, are followers of the Gandhian view of *satyagraha*, and therefore we can say that the Chipko movement was based on the ideology of nonviolence.

Tomalin's second point is that religious environmentalism is grounded in a Western tradition of environment and religion. Tomalin refers to Baviskar's writing about her work with Sangath, a local organization in Madhya Pradesh, which emerged with its concern of land rights, forest management, and funding to reach the local tribal leaders. Sangath leaders later participated in the NBA. Baviskar observed that a conflict gradually arose between the middle class and the *adivasis* (scheduled tribes). The middle-class people believed that sustainability was the way to address the situation and that it was best for the *adivasis* to remain close to nature. The *adivasis*, on the other hand, wanted control over the resources outside of the forest and be in control. The tribal leaders considered it arrogance when the middle-class people wanted their own economic security due to their literacy and imposed their ideas on the tribal leaders (Baviskar, 1997; Tomalin, 2002).

Tomalin claims that it was for pragmatic reasons that religion entered the Chipko movement. 'Those involved in the movement were Hindu, of course, but the movement was a struggle for land rights rather than an expression of religious values aimed at protecting the forest from destruction'. She concludes that 'Religion in India has not become privatized nor society secularized, and therefore the idea of the reemergence of religiosity into public life bears little relevance to the Indian context' (Tomalin, 2002, p. 25). In response to this, it has been argued above that it is not fair to put Hindu religious environmentalism in the same category as Western religious environmentalism, as they are two very different things. Hindu religious environmentalism is of a different category due to its complexity, and it is legitimate for Indian culture, as demonstrated above.

6.7 Influence of Hinduism and Other Literature on Gandhi

The influences that shaped Gandhi's thought were the teachings of the *Isha Upanishad*, Patanjali's *Yogasutras*, and the *Bhagavad Gita* as well as John Ruskin's book, *Unto the Last*. These influences fueled his desire to lead a life of frugality and develop ideas that foster 'sustainable development' (Khoshoo, 1995, pp. 1-2; Nimbalkar, 2017, p. 171).

Gandhi thought so highly of the *Isha Upanishad* that he remarked, 'If all the Upanishads and all the other scriptures happened all of a sudden to be reduced to ashes, and if only the first verse in the *Isha Upanishad* were left in the memory of the Hindus, Hinduism would live forever' (as cited in Easwaran, 2007, p. 53). The first verse of the *Isha Upanishad* is:

> The Lord is enshrined in the hearts of all.
> The Lord is the supreme Reality.
> Rejoice in Him through renunciation.
> Covet nothing. All belongs to the Lord. (Easwaran, 2007, p. 57)

Gandhi held an absolute commitment to truth and believed that by serving other human beings, one is serving God, as divinity is present in every human being. This is a fundamental principle of Gandhian *satyagraha* or truth-force.

From Patanjali's *Yogasutras*, the ethical guidelines of *yamas* (restraints) and *niyamas* (disciplines) were followed very strictly by Gandhi. The five *yamas* are nonviolence, truth, avoiding the use of materials obtained by illegitimate means, celibacy, and not accumulating wealth beyond one's needs. The *niyamas* of purification and contentment were also followed; cleanliness (*saucha*) refers to removing the negative things that affect our state of mind and body in order to achieve spiritual liberation. Contentment (*santosha*) refers to being content with the essentials of life and not going after material things but instead realizing that material gains are not long-lasting (Nimbalkar, 2017).

Another scripture that influenced Gandhi significantly was the *Bhagavad Gita*. In his interpretation of the *Gita*, Gandhi understood this scripture to be an allegory in which the battlefield is the soul, Arjuna is a person's higher impulses struggling against evil, and Krishna – the incarnation of God Vishnu – is the Dweller trying to steer Arjuna toward the right path of doing one's duty without attachment. 'The divine author has used a historical incident for inculcating the lesson of doing one's duty even at the peril of one's life. It inculcates performance of duty irrespective of the consequences, for we mortals, limited by our physical frames, are incapable of controlling actions save our own...The Gita distinguishes between the powers of light and darkness and demonstrates their incompatibility' (Gandhi, 1946-47, p. 127).

For Gandhi, the important lesson of the *Gita* is that any attachment to our ego is an illusion. The concerns of our body – which sees, smells, tastes, feels, and hears – are not real. The real self is what is called *Atman* (the self), which remains after our body dies and becomes associated with another body at some point. This is the reincarnation of the soul that takes place according to *karma* or action done. The *Gita* proposes three different ways of attaining *Atman:* the way of knowledge (*jnana yoga*), the way of devotion (*bhakti yoga*), and the way of action (*karma yoga*). Gandhi, of course, chose the way of action in his life, and called the *Gita* 'Mother', which nourished his spirit. The goal of *karma yoga* is to take action by doing one's duty for a just cause without thinking about what one could get out of it. This does not mean that the person who follows *karma yoga* should be indifferent or act out of frustration. It is just the opposite – one should act in the way of truth without brooding over the results. That is the requirement of the *Gita*. Hence, Gandhi wrote, 'He who is ever brooding over results often loses nerve in the performance of his duty. He becomes impatient and then gives vent to senses; he is ever distracted, he says goodbye to all scruples, everything is right in his estimation, and he therefore resorts to means fair and foul to attain his end' (Gandhi, 1946-47, p. 30).

Gandhi read all the original scriptures of other religions- – Christianity, Judaism, and lslam. He was greatly influenced by the Bible's 'Sermon on the Mount' and the teaching of 'If someone slaps you on the right cheek, turn and offer him your left' (Matthew 5:39 Wikipedia). He was also influenced by John Ruskin's book on the economy *Unto This Last*, and in 1908 Gandhi translated Ruskin's book into Gujarati under the title *Sarvodaya* (*Well-Being of All*). Gandhi used *sarvodaya* as the ideal for his political viewpoint. *Sarvodaya* is based on the following three principles: 1) that the good of the individual is contained in the good of all; 2) that a lawyer's work has the same value as the barber's in as much as all have the same right of earning their livelihood from their work; and 3) that a life of labour – that is, the life of the tiller of the soil and the handicraftsman – is the life worth living (Rathi, n.d.).

Gandhi also read Henry David Thoreau's works, taking his idea of civil disobedience and adding a spiritual component to it and calling it *satyagraha* to point out that it is different from civil disobedience, which does not necessarily have a religious component (Gandhi, 1983).

Gandhi was also greatly influenced by Leo Tolstoy's book *The Kingdom of God is Within You*. Gandhi had corresponded with Tolstoy from 1909 until Tolstoy's death in 1910. Tolstoy claimed that the path of violence has been taken as a guiding principle for survival, yet the path of violence is incompatible with the way of love. Tolstoy's philosophy of simplicity and his view that the way of love is the only law towards unity inspired Gandhi. As Tolstoy defended the rights of peasants, Gandhi named one of his farms in South Africa 'Tolstoy Farm' because it was self-sufficient and practiced manual labour to achieve the ideals of truth, non-violence, non-possession, and love.

6.8 Conclusion

Profound concern with nature is part and parcel of the diverse Hindu religious worldview, but the scope and extent of this influence is difficult to assess, as shown in this chapter, because of the complexity of the tradition itself and the specific socioeconomic problems of contemporary Indian citizens. The implicit value of the sacredness of nature in Hinduism ought to be recovered and celebrated. It should not simply be the case that this value is called upon only when one is dealing with poverty, displacement, and socioeconomic injustice. Inequality and socioeconomic injustice are the symptoms of a lost or forgotten value that capitalist development and consumerism have managed to submerge and preclude.

This chapter has provided a brief historical background of the Hindu religious periods, analyzing the issue of how Hindu *dharma,* which has both an ethical and a religious base, is different from the discipline of ethics in the West. Hinduism is the oldest religion in the world. The foundation of Hindu philosophy is metaphysical, attempting to understand reality as a guide to life. This chapter also examined the relation between Hindu *dharma,* ecology, and sustainable development by demonstrating the connections among the three. The question that arises from this is what things can be followed not only by Indians but also globally to address the ecological crises we face today. It has been shown that if one's approach is based on the day-to-day lives that Indians follow with the rituals and social intentions of Hinduism, it is possible to address the issue of ecological crisis and sustainability, which are crucial issues at present. Indeed, the present ecological crisis demands that the sacredness of nature celebrated in Hindu scriptures be integrated with the daily rituals, social behaviour, and environmental awareness of Indians in order to bring about a sustainable universe.

Works Cited

Agarwal, Anil (2000). 'Can Hindu Beliefs and Values Help India Meet its Ecological Crisis'?. In Christopher Key Chapple and Mary Evelyn Tucker (eds.), *Hinduism and Ecology: The Intersection of Earth, Sky, and Water.* Cambridge, MA: Harvard University Press, 165-182.

Alley, Kelly D. (1998). 'Idioms of Degeneracy: Assessing Ganga's Purity and Pollution'. In Lance E. Nelson (ed.), *Purifying the Earthly Body of God: Religion & Ecology in Hindu India.* Albany, NY: State University of New York Press, 297-330.

Baviskar, Amita (1997). 'Tribal Politics and Discourses of Environmentalism'. *Contributions to Indian Sociology,* 31, 2, 195-224.

The Bhagavad-Gita: Krishna's Counsel in Time of War (1986). Translated by Barbara Stoller Miller. New York: Bantam Books.

Berry, Thomas (2000). *The Great Work: Our Way into the Future.* New York: Bell Tower, Reprint Edition.

Bilimoria, Purushottama (1998). 'Environmental Ethics of Indian Religious Traditions'. In David E. Cooper and Joy A. Palmer (eds.), *Spirit of the Environment: Religion, Value and Environmental Concern.* London & New York: Routledge, 1-14.

Calicott, J. Baird (1997). *Earth's Insights: A Multicultural Survey of Ecological Ethics from the Mediterranean Basin to the Australian Outback*. Berkeley, CA: University of California Press.

Chaudhuri, Nirad (1979). *Hinduism: A Religion to Live By*. New York: Oxford University Press.

Creel, Austin B. (1977). *Dharma in Hindu Ethics*. India: South Asia Books.

Dhand, Arti (2002). 'The Dharma of Ethics: The Ethics of Dharma: Quizzing the Ideals of Hinduism'. *The Journal of Religious Ethics*, 30, 3, 347-372.

Easwaran, Eknath (2007). Second Edition, *The Upanishads, Translated for the Modern Reader*. CA: Nilgiri Press.

Fisher, R.J. (1997). *If Rain Doesn't Come: An Anthropological Study of Drought and Human Ecology in Western Rajasthan*. New Delhi: Manohar.

Gandhi, Mohandas K. (1983). *Autobiography: The Story of My Experiments with Truth*. New York: Dover Publications Inc.

— (1908). *Sarvodaya*. Ahmedabad: Navjivan Trust, 2010 paperback.

— (1946-47). *Gita: The Mother*. Edited by Jag Parvesh Chander, fourth edition. Indian Printing Works.

Guha, Ramachandra, and Juan Martinez-Alier (1997). *Varieties of Environmentalism: Essays North and South*. London: Earthscan.

Haberman, David (2006). *River of Love in an Age of Pollution; The Yamuna River of Northern India*. Berkeley, CA: University of California Press.

Hamilton, Clive (2008). *The Freedom Paradox: Towards a Post-Secular Ethics*. Crows Nest, New South Wales: Allen & Unwin.

—, and Richard Denniss (2005). *Affluenza: When Too Much is Never Enough*. Crows Nest, New South Wales: Allen & Unwin.

Jain, Pankaj (2011). *Dharma and Ecology of Hindu Communities: Sustenance and Sustainability*. Milton Park and New York: Routledge.

— (2010). 'Jainism, Dharma, and Environmental Ethics'. *Union Seminary Quarterly Review*, 63, 121-135.

James, George Alfred (2014). 'Environment and Environmental Philosophy in India'. In J. Baird Calicott and James McRae (eds.), *Environmental Philosophy in Asian Traditions of Thought*. Albany, NY: State University of New York Press, 3-23.

Jhingram, Saral (1989). *Aspects of Hindu Morality*. Delhi: Motilal Banarsidass.

Kent, Eliza (2016). 'Hinduism and Environmentalism in Modern India'. In Brian A. Hatcher (ed.), *Hinduism in the Modern World*. New York: Routledge, 290-308.

Khoshoo, T.N. (1995). *Mahatma Gandhi: An Apostle of Applied Human Ecology*. New Delhi: Tata Energy Research Institute.

Kuppuswamy, B. (1977). *Dharma and Society: A Study in Social Values*. Delhi: Macmillan Co.

Mahabharata, Shantiparva (1927-1954). Poona, India: Bhandarkar Oriental Research Institute.

Matthew 5:39 Wikipedia.

Mohanty, J.N. (2007). 'Dharma, Imperatives and Tradition: Toward an Indian Theory of Moral Action'. In Purushottama Bilimoria, Joseph Prabhu, and Renuka Sharma (eds.), *Indian Ethics: Classical Traditions and Contemporary Challenges*, Volume 1. UK: Ashgate Publishing, 57-78.

Nadkarni, M.V. (2011). *Ethics for Our Times: Essays in Gandhian Perspective*. India: Oxford University Press.

Narayanan, Vasudha (2001). 'Water, Wood, and Wisdom: Ecological Perspectives from the Hindu Traditions'. *Daedelus*, 4, 179-206.

Narayanan, Yamini (2013). 'Inspiring Sustainability beyond Sustainability: Sustainable Development and the Ultimate Hindu Purpose'. *Nature and Culture*, 8, 3, Special Symposium on 'Religious Innovation for a Sustainable Future: Perspectives from Norway, Ghana, China, and India', 301-323.

Nelson, Lance E. (1998). 'The Dualism of Nondualism: Advaita Vedanta and the Irrelevance of Nature'. In Lance E. Nelson (ed.), *Purifying the Earthly Body of God: Religion and Ecology in Hindu India*. Albany, NY: State University of New York Press, 61-88.

Nimbalkar, Namita (2017). 'The Traditional Hindu Perspective on Environment and M.K. Gandhi's Standpoint'. In Hiralal Kalpana (ed.), *Global Hindu Diaspora: Historical and Contemporary Perspectives*. London and New York: Routledge, 163-177.

Perrett, Roy W. (1998). *Hindu Ethics: A Philosophical Study*. University of Hawaii Press.

Prasad, Kiran (2013). 'Environmental Communication from the Fringes to Mainstream: Creating a Paradigm Shift in Sustainable Development'. In Jan Servaes (ed.), *Sustainable Development and Green Communication: African and Asian Perspectives*. UK: Palgrave Macmillan, 95-109.

Radhakrishnan, Sarvapalli (2005). *Bhagavad Gita*, New Delhi: Harper Collins.

Rathi, Shubhangi. https://www.mkgandhi.org/articles/gandhi_sarvodaya.html (accessed 20 February 2020).

Ruskin, John (1910). *Unto this Last*. London & Glasgow: Collins' Clear Type Press.

Sharma, V.D. (1999). 'Bishnois: An Ecoreligion'. *The Hindu Survey of the Environment*. Chennai: The Hindu.

Steir-Younis, Alfredo (2001). 'The Spiritual Imperatives of this Millenium: Healing Humanity'. http://www.worldpeacecongress.net/en/2004/speakers/alfredo.htm (accessed 7 June 2020).

Tolstoy, Leo (1984) *The Kingdom of God is Within You: Christianity Not as a Mystic Religion but as a New Theory of Life*. Translated by Constance Garnett. Lincoln, NE: University of Nebraska Press.

Tomalin, Emma (2002). 'The Limitations of Religious Environmentalism for India'. *Worldviews*, 6, 1, 12-30.

Weiss, Richard S. (2019). *The Emergence of Modern Hinduism: Religion on the Margins of Colonialism*. Oakland, CA: University of California Press.

Conclusion

The Symbiosis of Natural Resources and Local Needs

Abstract

The conclusion reiterates the three main messages of this book: that the human and the environmental world are interrelated; that economic and ecological development are not mutually exclusive; and that there is a relationship between the local and the global. Global activism is valuable, as it helps to bring to the forefront the voices of the disadvantaged. From a moral point of view, the well-being of all humans – both the advantaged and the disadvantaged – are equally important. The Chipko, NBA, and Navdanya movements have made a global impact by taking the discourse on economic development to a new level that includes the voices of the powerless and the centrality of environmental sustainability. In order to make progress towards global environmental sustainability, a paradigm shift must be made in democratic societies.

Keywords: global environmental theory, feminist care ethics, ethics of nonviolence

As India goes through a sweeping process of economic transformation, environmental sustainability concerns must be at the forefront of its development discourse. India's food grain production has increased considerably from the 1960s to the 1980s, and its economy has developed significantly in the 1990s and 2000s. One might think that, with its robust economic development and self-sufficient food production, all segments of the Indian population have benefited and all issues related to food sovereignty have been resolved. That is not the case today.

The chapters of this book have presented perspectives on the economic development discourse as they took shape within the Chipko movement (1970s), the Narmada Bachao Andolan (1980s and 1990s), and the contemporary

Mallick, Krishna, *Environmental Movements of India: Chipko, Narmada Bachao Andolan, Navdanya*. Taylor & Francis Group, 2021.
DOI: 10.5117/9789462984431_CONCL

Navdanya movement. Social justice is at the core of each of these three movements. Social equity is an important component of social justice. In an equitable society, there should not be any discrimination that prevents people from functioning socially, politically, and economically. Unfortunately, this is not the case in the real world.

By social justice, we generally mean how both the good and bad aspects of life are distributed among the members of a society. Social justice is related to 'distributive justice', meaning how the goods and services of a society are distributed. Some of the themes of social justice involve equity, human rights, environmental justice, sustainability, and globalization. These three movements are considered as social justice movements, but they are also related to environmental justice, since the social injustices are in fact found within environmental conditions. Social and environmental justice include not only equity, recognition, and participation; they also encompass concerns related to the basic needs and functioning of individuals and communities.

Each of the movements is characterized by a struggle with local issues related to the livelihood of the people living in the affected communities. What their accomplishments imply – or show explicitly – is that 'globalization from below' is the only way to resolve or remedy existing issues or at least to not make such issues any worse off than what they are at the moment. Focusing on 'local' needs is crucial to the approach of 'globalization from below'. Local activists are creating awareness of injustices being committed in relation to the environment as well as enabling changes that promote a more just society while making connections at the global level. Local activists focus on the mandate of the people they represent, protecting the rights of the disadvantaged and disempowered minorities by keeping communications open through technology.

I have demonstrated that a 'one-size-fits-all' approach cannot be effectively applied to environmental issues in every country of the world nor in any one part of India. The specific situations of those local communities that have been directly and negatively affected must be brought out into the open within their historical background. This leads to a profound understanding and recognition that movements are evolving within so-called 'progress in economic growth and technology'.

It is worth noting that the activists of these three movements have attempted to bring about structural change not only locally but also globally. Each of these movements has taken the development discourse to a different level by making authorities listen to the voices of those who are most negatively affected by actions taken by policymakers and big businesses not only in India but also in the West. The central questions that these movements pose are as follows:

1. How should humans belonging to a certain group/community treat other human beings belonging to another group/community or geographical region in addressing the social injustices that occur? Is it morally justified to appeal to what is natural as a justification of a particular action?
2. How can activism create a more democratic and diverse world in such a way that all peoples' voices matter?
3. How are the advantages and disadvantages of Indian society distributed among the different constituents of the community?
4. Is it morally right to take away the rights of one group of disadvantaged people – specifically their right to livelihood – for the benefit of the majority group, who are not necessarily disadvantaged?

Theoretical Views of the Global South

The three movements can also be examined through the four distinct theoretical lenses of global environmental theory, feminist care ethics, the capabilities approach, and the ethics of nonviolence.

Global Environmental Theory

Some scholars (for example, Wapner & Matthew, 2009) argue that if humans cannot treat other humans with dignity, then it is even more doubtful whether morality can flourish beyond the human world. Environmental ethical scholarship in the Western world has evolved around biocentrism and ecocentrism by being critical of anthropocentrism. Leopold's 'land ethic' view criticized anthropocentrism by arguing how moral considerations must be extended beyond the boundaries of humanity to include the land community. Later on, environmental philosophers like Naess (1973), Rolston (1975), and Singer (1975) followed up on Leopold, adding that not only the human world but also the non-human world has moral worth. Although an analysis of the relation between the human and non-human worlds has enormous significance in an international context, with regard to these three environmental movements, it is important to point out that much of the environmental harm caused involved the exploitation, abuse, or mistreatment of some human beings by others.

In the case of the Chipko movement, the livelihood of the women living in the Himalayan region was taken away when the lumber companies cut down the forest for profit. It can be acknowledged that the lumber companies did not do anything illegal, as they were following the colonial period's Forest Policy. Nevertheless, their actions ought to be judged as immoral insofar as they involved taking away the rights of these women, whose job was to collect the branches from the trees and bring them home to make fire and cook food for their children. In the case of the NBA, the rights of the indigenous people were alienated when they were forcibly displaced, which simultaneously took away their means of survival. In many cases, moreover, they were not compensated appropriately. Similarly, in the case of the Navdanya movement, the protests against the MNCs and TNCs were motivated by moral outrage, especially about patent rights being taken away from small farmers and about their right to livelihood being deprived. In each of these cases, the group or people in power took away the rights of the disempowered and the disenfranchised.

As Wapner and Matthew put it, 'environmental injustice arises because people tend to redirect, rather than resolve environmental dilemmas, conveying their burdens onto vulnerable others through the geographies of power. Such displacement takes place across both space and time' (2009, p. 208). Global environmental theory argues that biocentrism or ecocentrism as an environmental ethical theory does not have any value in the Global South, where people's primary concern is their own livelihood and not the preservation of wilderness.

Feminist Care Ethics

The Chipko movement demonstrates the application of the ethics of care by showing the interconnectedness between forests, women, and communities. This interconnectedness was symbolically expressed through the ritual of tree-hugging. Similarly, in the case of the NBA, the *adivasis* (indigenous people) who have been displaced without proper compensation have in some cases demonstrated the ethics of care by showing their relationship to nature as being integral to their living. To take away or undermine the *adivasis*' integral connection with the natural environment is morally degrading. The *adivasis* are unable to cope with the conditions that they have been forced to endure. And in the case of the Navdanya movement, the denial of the rights of the local small farmers is at the same time a failure to recognize that the local community is an integral part of the broader society.

The ethics of care highlights the affective dimension of morality, unlike more traditional ethical approaches that prioritize the rational aspect and consider the affective aspect of human nature as subordinate (Held, 2007). The ethics of care teaches us that moral agents are integrally related to other people and the nonhuman world. The field of feminist ethics of care has evolved by applying it to the development policy and sustainability discourses. It recommends that we act in a responsible and virtuous way towards others, including the future generation and every being in the ecosystem by positively influencing the fundamental connection between the ecological social relationships even in cases where their interests might overlap.

The Capabilities Approach

The capabilities approach, as developed by Amartya Sen (1999) and later on by Martha Nussbaum (2000), is used to evaluate different aspects of human well-being such as the inequality, poverty, and average well-being of the members of a group. The aim of the approach is to bring about social change. Its focus is on what people are able to do or who they are capable of becoming. In other words, the capabilities approach emphasizes human capabilities and cares about improving the quality of people's lives by removing the obstacles that prevent them from living more freely and securing the values that they find worth pursuing. The capabilities approach covers all dimensions of human well-being – economic, social, political, and cultural. It encompasses basic needs, social recognition, and economic and political rights. It has likewise provided a much broader theoretical framework for people to understand the demands of many environmental justice movements. A major offshoot of the capabilities approach is represented by the six core values of development ethics and is exemplified in the theoretical foundations of the NBA movement.

Ethics of Nonviolence

Each of the three movements has employed the practice of nonviolence, which is based on ethics. The ethics of nonviolence involves an absolute commitment to the principle of justice or truth and requires people to prepare themselves to go through the suffering inflicted by an antagonist without using any form of violence in thought, speech, or action. They must be prepared to not surrender until the antagonist realizes that what s/he is doing is unjust and subsequently changes his/her unjust behaviour.

As I have stated in *An Anthology of Nonviolence: Historical and Contemporary Voices* (2002), the members of the Chipko movement followed the Gandhian method of *satyagraha* by confronting the Indian government and making it realize that what the lumber companies were doing was unjust. Their actions led the government to ban tree-cutting in this Himalayan region for 15 years. Some scholars, such as Robert Holmes (2013), have argued that the ethics of nonviolence relates to the Western theory of virtue ethics. I would argue that the Gandhian method has not only the component of character-building and integrity but also an equally or even more important component of making an absolute commitment to the principle of justice with an eye toward changing ongoing injustices. As I have also demonstrated throughout this book, the ethics of nonviolence, which originated from many Hindu scriptures, has had a profound influence on the members of the Chipko, NBA, and Navdanya movements, and its principles have been implemented in practice.

The main themes that appear in each of the chapters of this book are sustainability, feminism, nonviolence, globalization, people's power, and understanding the context under which the three grassroots movements have taken place in India, the world's largest democracy. This book has conveyed three core messages. First, it has emphasized the interrelatedness of the human and environmental world, noting that what is done to one has an inevitable effect on the other. Second, economic development and ecological development are not mutually exclusive. And the third core message is that there is a relationship between the local and the global. The local situations are very different, but they can have a global impact at different levels.

My contribution in writing this book has been to expand the understanding of global activism by re-visioning a world in which a paradigm shift needs to be made towards the emotional well-being of individuals, groups, and communities both at the micro and macro levels. In addition, I propose integrating the sacredness of nature as found in many Hindu scriptures with the social behaviour of people in India by raising environmental awareness so that the environment becomes more sustainable not only in India but also globally.

Those wishing to explore this subject matter further might ponder whether this paradigm shift is going to take place. What is the direction that environmental activism is taking? With protest movements being so popular at present in all the democratic countries of the world, does and will environmental activism continue to have a local as well as global impact, or is it just a temporary phase without any principled commitment? What role has Western environmentalism played in influencing the environmentalism of the Global South? What is the common language of both sides of the world?

Works Cited

Held, Virginia (2007). *The Ethics of Care: Personal, Political & Global.* New York: Oxford University Press.

Holmes, Robert F. (2013). 'The Morality of Nonviolence'. In Predrag Cicovacki (ed.), *The Ethics of Nonviolence: Essays by Robert Holmes.* New York: Bloomsbury Academic, 169-182.

Mallick, Krishna (2002). 'Gandhian Satyagraha and Chipko Movement of India'. In Krishna Mallick & Doris Hunter (eds.), *An Anthology of Nonviolence: Historical and Contemporary Voices,* Greenwood Press, 279-285.

Naess, Arne (1973). 'The Shallow and The Deep: Long-Range Ecology Movements'. *Inquiry,* 16, 95-100.

Nussbaum, Martha (2000). *Women and Human Development: The Capabilities Approach.* Cambridge: Cambridge University Press.

Rolston, Holmes (1975). 'Is There an Ecological Ethic?' *Ethics: An International Journal of Social and Political Philosophy,* 85, 93-109.

Sen, Amartya (1999). *Development as Freedom.* New York: Knopf.

Singer, Peter (1975). *Animal Liberation: A New Ethics for our Treatment of Animals.* New York: Random House.

Wapner, Paul, & Richard A. Matthew (2009). 'The Humanity of Global Environmental Ethics'. *The Journal of Environment and Development,* 18, 2, 203-232.

Index

Access and Benefit Sharing (ABS) 78
  see also Nagoya Protocol
achit 160
active resistance 136
adivasis
  as scheduled tribes 167
  displacement of, by NBA 60
  see also Feminist Care Ethics
Advaita Vedanta 143, 146, 165, 173
Agreement on Agriculture 85
agribusiness 99, 101
agricultural diversity and productivity 83
Agricultural Research Institute of India and Raichur 108
agriculture
  and food sovereignty 90
  organic 31
  subsistence 41, 65
  sustainable vs. industrial 83-88, 92-107, 111-114
Agroecology 112
ahimsa 18, 146, 153
Alakananda valley
  flood (1970) 38
  government auction of Reni trees 43
Amendment of the Biosafety Rules (1989) 89
  protection of biodiversity 90
Anti-GM movements 87-109
Arendt, Hannah 90, 110
Arjuna 150-151, 157, 160, 169
artha 145, 154-155
Atman 149, 154, 157, 169

Bacilus thuringiensis (Bt) cotton
  allegations of monopolization 84-85
  patent 107
  protests against 87
  see also Monsanto
Bacon, Francis 104
Badwani, site of protest in Madhya Pradesh 66
Bahuguna, Sunderlal
  as co-leader of Chipko movement 30, 37
  nonviolent activism 43-46
  political savvy 54-56
Bajgis, musicians of Chipko 39
Bechtel
  and control over water 99
Bhagavad Gita
  as read at Chipko protests 39
  influence on Gandhi 27, 168-169
  moral truths 30, 150
  philosophy of 155, 157, 160
  see also Mahabarata
Bhakti movement 147
Bhatt, Chandi Prasad
  as co-leader of Chipko movement 29
  awards and admiration 45-46
  founding of workers' cooperative 42
  suggestion to hug trees 38, 43
  training of activists 46
Bhil communities 162-163
Bhoomi Festival (2012) 92
bij Swaraj 96, 105
Bija Vidyapeeth (Earth University) 93
bilva tree 166
biocentrism 177-178
biocide 158
biodiverse organic farming 92-93
biodiversity 93-94, 99-100, 102, 106-109
biopiracy 89
biospherical egalitarianism 116
biotechnology
  agricultural 84, 106
  debated cause of farmer suicides 84-86
  destructive effects of 96
Bishnois 161-162
Bolivian activists 102
Brahmanical tradition 163-164
Brahmin 117
British Raj 19, 23, 33, 41
British Rule (1757-1947) 23, 33, 145
  see also East India Company
Bt brinjal 107
Buddha 44
Buddhism 144-146

capitalism
  European 41
  international 100
Central Indian government 65
Chamoli Workers' Cooperative 42
Chipko (Hug the Trees) Movement 17-20, 23-30, 32
  as an act of survival 38
  ecofeminist roots 133-135
  debate over success 49-55
  formation of movement 40
  moral implications 116-119, 133-137
  role of women 46-48
  use of ancient Indian texts 39
chit 160
civil disobedience 83-84, 99
climate change
  as a social challenge 93
  biodiverse organic farming 93-95
  drought-resistant varieties 107-109
  evidence of, in Himalayas 93
  Gene Campaign for farmers' rights 88
  Navdanya campaign in Garwhal 93-94

Coca-Cola 103
common *dharma* 150-152
Conference of the Parties (2009) 105
contentment 153, 169
  see also santosha
corporate greed 91, 136
corporations, transnational 18, 83, 90-91, 99-102, 109
cosmopolitan federalism 97
Cremate Monsanto campaign (aka Quit India) 87-88
  see also KRRS
crop rotation 40

dam building 19-20, 25
  anti-dam sentiments and displacement 52-54
  economics and rationale 60-62
  gendered nature of 70, 77
  moral implications 123-129
Dandi march (1930) 28
darshan 145
Darwin, Charles 104-105
Dasholi Gram Swarajya Sangha (DGSS) 42-45, 51
Dashuli Village Self-Help Association 29
  see also Bhatt, Chandi Prasad
deep ecology 116-119
deforestation 17-18
  by Britain (1860) 24
  commercial and industrial 41, 55
  impact on women 47, 135
  in present-day India 165
  protests against 38-41
Delta and Pine Land (acquired by Monsanto in 2006) 85
Department of Biotechnology (DBT) 85
de-sacralization of nature 103
desalination plants 124
Descartes, Rene 30, 104-105
detachment 157
developing countries 32, 84, 109, 120, 131-132
development
  as detrimental to humans 118-121
  economic considerations of 60-64, 86
  exploitation of tribal communities 26
  in relation to Hindu philosophy 153-158
  integrated with ethics 71, 74
  of agricultural biotechnology 84
  of rural areas 92
  of self for global sustainability 156
  post-colonial 66
  secular, and social justice 49-51
  sustainable alternatives and approaches 46, 76, 86, 100, 110
displacement
  as a concern of Development Ethics 122, 125, 127-128
  by large infrastructure projects 66
  of *adivasis* by NBA 60, 67
  women's roles in resettlement 72
Development Ethics 121-130
Devi, Gaura 47-48
dharma 31
  as ethical duty 150-153
  and reverence for nature 161-163
  compared to western thought 147-150
  connection to sustainable development 157-160
  historical background 145
  principles of behavior 154-155
dharma-shastras 146-149
dharmasutras 146
displaced populations 59, 62, 64-66, 69-71, 78
  impact on women 72-75
  see also oustees
distributive justice 31, 176
drip irrigation 107
drought-resistant sugarcane 106
dualism 131-132
  and non-dualism 164-165
  patriarchal 134-135

Earth Charter Commission (2000) 104
Earth Democracy 31, 103-105
Earth University 93, 95
ecocentrism 177-178
ecofeminism 72, 109, 121, 131-136
ecofeminist ethics 131, 135
ecofeminist movements 116
ecofeminists 73, 103, 131, 133-136
ecological and embodied debt 131
ecological crisis 46, 118, 144, 156, 158, 163
ecology 49-50, 89, 103, 131, 136
  and relation to Hindu texts 153, 159, 161-162, 171
economic development 24, 32, 47, 49-50, 61, 67, 78
economic justice 59, 137, 154
eco-religion 161
egalitarianism 116
  lack of, in Chipko households 133
Emancipatory justice 70
  see also Simon Company
'empty belly environmentalism' 166
enlightenment 155-156
entrepreneurship 95
environmental crisis 143-144, 159-160
environmental justice 20, 70, 176
  gendered aspects of 130-131
environmental movements
  and distributive justice 31-32
  anti-GM 87-90
  Gandhian 60
  in present-day India 94
environmentalism 33-34, 46
  in Narmada valley 72
  in Global South 101
  influenced by Gandhi 117-119
  religious 143-144, 161, 163, 167-168

equity 32-33, 61, 74-75, 125-126, 176
ethical consumption 100
ethics of care 110, 135
ethics of nonviolence 179-180

Fadnaviss, Devendra 107
farmers 24, 61
  and climate change 106-109
  biodiverse organic farming 92-93
  Navdanya cooperatives 89-90, 95-96, 102
  protests against GMOs 87-88
  suicides 84
  well-being 107-108
Feminist Care Ethics 176, 178
food rights 89
food security 83-84, 88, 91-92
food sovereignty 90-91, 96-98
Forest Acts 23, 26
Forest Act of 1927 26, 119
Forest Conservation Act (1980) 42
Forest Department 26, 38, 41-42, 45, 51, 55
forestry
  agro- 45
  British colonial and scientific 25, 41
  commercial 37, 41
  community management of (*van panchayat*) 26, 42, 47
  subsistence 73
four pillars of nonviolence 96
  see also Swaraj, Sarvodaya, Swadeshi, Satyagraha
Fourth World Women's Conference in Beijing (1995) 74

Gaia Foundation 101
Gandhi, Indira 54, 56
  Stockholm remarks (1972) 72-73
Gandhi, Mohandas K. (*Mahatma*)
  as patron saint of environmentalists 115
  biography 27
  Dandi salt march 28
  literary influences 27, 168-170
  philosophy of nonviolence 115-117
  philosophy of truth (*satyagraha*) 27-29, 119
Ganges River 164
Garwhal campaign 93
Garhwal and Kumaon 42, 53
Gaura Devi 47-48
gender equality 74
Gene Campaign 88, 106
genetic engineering 89, 107
Genetic Engineering Approval Committee (GEAC) 107
genetic use restriction technologies (sterile seed production) 85
genetically modified (GM) organism (GMO) 84-86
  opposition to 87-90
  moral dilemma with regard to climate crisis 106-109
geocide 158
Gita 27, 30, 43
  three paths to salvation 149-151
  guidance for detachment 157
  influence on Gandhi 166-169
  see also Bhagavad Gita
globalism 120-121, 136
globalization 119-121, 132, 136
Global Environmental Theory 177-178
Global Forum on Food Sovereignty (2007) 91
Global North 76, 101, 120
Global South 76, 99, 101 121, 136-137, 177-178
GM crops 106-107
GM mustard 107-109
GM seeds 87
  U.S. origin of 88
God Vishnu 27, 160, 166, 169
grassroots movements 131, 180
gravity-well systems 124
Green Revolution 83, 94, 96, 99, 103
Gujarat 27
  as drought stricken region 30
  as state impacted by the Narmada dam 60, 64-65
  oustee compensation and rehabilitation 69-70
  state funding 124
gunas 15, 151
gurus 161, 163

Harijan, Gandhi's journal 116
Haryana, home to Bishnois 161
herbicides 99
Hind Swaraj 27, 118
Hindu
  dharmic civilization 27
  ethics 143-145, 147-150
  extremist groups 30
  literature 149-153
  philosophy 157, 160, 167, 171
  religious periods 145-147
  renaissance 147, 164
Hinduism
  and ecology 153-154
  as a sustainable system 144-145
  codes of conduct 150
  influence on environmental movements 29-31
  metaphysical dimensions of 143-146, 151, 154, 156-157, 164, 171
hybrid seeds 85, 107-109
hydroelectricity 63, 129

Indian Council of Agricultural Research (ICAR) 106-107
immortality 156-157
independence (1947) 19
  agricultural demands 83
  continued exploitation of forests 25-27, 41, 49

impacts on tribal areas 74
License Raj 95
Independent Review of SSP 62, 67
Indian environmental movements
contrasted with U.S. 33
social justice concerns of 31-32
Indian Forest Act of 1865 24
Indian Forest Act of 1878 24
and rise of countryside protests 25
Indian government
Central Water and Power Commission 65
Constitution 26, 66
Ministry of Environment and Forests 65, 73, 130
National Agricultural Policy (NAP) 86
Official Secrets Act 127
Planning Commission 65
water usage 106-107
Indian Patent Act 85
Indian philosophy 103, 144, 148-158
indigenous farmers and civil disobedience 84
indigenous populations
as movement participants 45-47
knowledge and culture of 88-89
rights of 34, 41, 60, 121, 126, 128-129
subsistence lifestyle of 65
see also adivasis
industrial agriculture 83, 104, 105
industrial deforestation 167
interconnectedness 37, 83
interdependence 116, 120
International Commission on Future of Food 101
international development 63
and biotechnology 84
International Monetary Fund 99-100, 155
International Peasant Movement 90
Inter-State Water Disputes Act 64
irrigation 39, 60-63
Isha Upanishad 168

Jainism 27
heterodox or atheistic system 144-146
jnana yoga 150, 157, 169
justice
ecological 97
economic 59
emancipatory 70
environmental 70, 130-131, 137
for oustees 125
for a sustainable future 103-105
gender 70-72
social 32-33, 39, 49, 56
Justice, Sustainability and Peace (2016) 103
see also Vandana, Shiva

kama 145, 154-156
Kantian ethics 151
karma yoga 150, 157, 162, 169
Karnataka 83, 87
as drought-stricken region 106
Karnataka Rajya Raitha Sangha (KRRS) 87
khadi 97
King Dasaratha of Kosala 152
see also Ramayana
Krishna, Lord 160, 163-165, 169
kshatriyas 117, 145
kuladharma 150
Kumaon 38, 41-42
alcohol use among men 53
Kumaon and Garhwal
protests of timber exports (1921) 25-26
Kumaon Forest Grievances Committee 26
Kumar, Virendra 44

lack of water 161
Land Acquisition Act (1894) 24, 66
land rights 167-168
land sovereignty 106
large dams
and worsening climate change 130
drawbacks and alternatives to 62
gendered dimensions of 74
global statistics and financing 61
social costs and benefits 74-75, 126-130
Laws of Manu 161
legislative acts 24
Lele, Sharachandra 94
liberation ecology 49-50
License Raj 95
'little tradition' 164
local communities 24-25, 54, 70, 78, 94, 102-104, 128, 176
Lord Krishna 160, 163-165, 169

Madhya Pradesh, state affected by Narmada dam 59-60
conflict with neighboring states 64-65
Sept. 1989 protest in Badwani 66
resettlement of oustees in city slums 125
Mahabharata 146, 150
and *Bhagavad Gita* 151
codes of common *dharma* 150
on wealth 155
Maharashtra
affected by Narmada dam 59-60, 127-128
home to Swadhyayees 162
temporary ban on Bt cotton 85
water shortage due to sugar cane crop 106-107
Maharshi, Ramana 147
Mahatma Gandhi 27, 42, 45, 48, 115, 147
see also Gandhi, Mohandas K.
Mahyco 85, 107
Mahyco-Monsanto Biotech (MMB) 85, 89
maldevelopment 123-124, 128
managerial environmentalism 33
Mandal village 51
see also Chipko movement

Manu, law giver 160-161
masculinist worldview 70-71
materialist ecofeminism 110, 121, 135-136
meaning of life 158
meaninglessness 156
metaphysical foundations of Hinduism 143-146, 151, 154-157, 164
Mimamsa 144
Ministry of Environment and Forests 65, 73, 130
Modi government 107-109
moksha 146, 148-149, 151, 154-156
monocropping 83-84, 96
monopolization 84-85, 98-99
Monsanto
  and monopoly through policy 98-99
  buying into partnerships 84-85, 89
monuments, sacred 129
moral authority 165
moral code 144, 148, 153-158
  see also dharma
moral dilemma
  of GM drought-resistant seeds 108
  of Arjuna in the *Gita* 150
moral worth of animals in the Global South 177
  see also Global Environmental Theory
morality
  and ethics of care 179
  in particular, compared to universal 149
  lack of, in neoliberalism 97
  of Bhakti movement to reject caste assumptions 147
  of self, to achieve *moksha* 155-156
  relation to water access 161-162
  social 154
Morse report on World Bank funding 67
Mother Ganga 163
multinational corporations (MNCs) 90-91, 98, 178
  Navdanya dilemma 98
  on wastefulness 121
  protests motivated by moral outrage 178

Naess, Arne 117-119
  see also deep ecology
Nagoya Protocol (2014) 78
Narmada Bachao Andolan (NBA)
  alternative recommendations 62
  as an example of Gandhian principles 30
  consolidation of NGOs 65-66
  critique of the state 60
  economic justice 59-60
  gender inequalities 68-69, 71-73
  halt and resumption in construction (1995) 76
  human rights abuses 122, 126
  international recognition (1980s) 76-77
  petition to Supreme Court (1994) 66-67
  protests against displacement 60, 66
  success factors 75
Narmada Control Authority (NCA) 65
Narmada valley 30, 65, 71
Narmada Valley Development Authority (NVDA) 65
Narmada Valley Development Project (NVDP) 76, 121
  as maldevelopment 122-125
  complaints against 126-130
Narmada Water Disputes Tribunal (NWDT) 'Award' 64-65
National Alliance of People's Movements (NAPM) 77
National Agricultural Policy 86
National Committee on Environmental Planning and Coordination 73
National Programme on Organic Production (NPOP) 92
native varieties 102
Navdanya (Nine Seeds) movement
  as an example of Gandhian principles 31, 91
  and reliance on corporate donations 98
  biodiversity conservation 89-90
  climate science 93-94
  conscientious objection and civil disobedience 96-98
  entrepreneurship 95-96
  food sovereignty 90-91
  grassroots organization 94
  history and operations 89, 92-94, 102
  international notoriety 101
  success factors 99
  opposition to GM crops (1987) 87-88
  philosophy of 101
  see also Earth Democracy
  see also Shiva, Vandana
Nehru, Jawaharlal 64
neo-Hinduism 164
neoliberal capitalist development 73
  gendered dimensions of 74
  see also multinational corporations (MNCs)
neoliberal economics 121
neoliberal globalization 87-88, 120
neoliberalism and immorality 97
  U.S. origin of GM seeds 88
Newton, Isaac 104-105
niyamas 153, 168-169
nondualism, qualified 160
non-governmental organizations (NGOs) 46, 66, 87, 92, 98, 127
non-maleficence as a development goal 123-124
nonviolence 28, 31
  ascetic vs. warrior 117
  ethics of 179-180
  four pillars of 96
  Gandhi's description of, in *Harijan* 115-117
  Naess' description of relation between all living beings 119

suffering to change hearts 116
see also satyagraha
nonviolent resistance 39
'red' and 'green' ideologies in NBA movement 60
Nussbaum, Martha 179
Nyaya 144
Nyéléni Forum 91

organic
agriculture 32, 88
biodiversity 93
connection among all living things 160
farming 86, 89, 92-93
entrepreneurship 92, 95-96
production 89, 92
seeds 83
oustees
and loss of cultural treasures 129
compensation only to men 71
psychological harm and precarious futures of 125

padayatra
Bahuguna's Himalayan march 43
Gandhi's Salt March 28
Patkar, Medha 59, 62, 70, 72-73, 75, 77
patriarchal dualism 134-135
Pawar, Sharad 106
people's power 144, 180
pesticides 83, 99, 105
philosophy, Indian
as compared to the West 147-150
interrelated with religion 144
metaphysical basis of 144
purusarthas 154
qualified nondualism 160, 165
qualities of being (*gunas*) 150
six orthodox schools 145-147
world as one family 103
see also Ramayana and *Mahabarata*
Plumwood, Val 103, 132
post-colonial development 66
poverty
and gender inequality 74, 130
of farmers and indigenous people 93, 121
resulting from timber extraction 55
'the greatest polluter' 73
Prakriti 146, 149, 151, 160, 164
preservationist view 30
principle of universalizability 147-148, 151
Priyadarshini Vrikshamitra Award 48
see also Devi, Guara
pro-dam as a 'masculinist' worldview 70
Punjab 161
Puranas 146, 159, 165
pure consciousness 152
purusarthas 154-155
Purusha 146, 150, 152
Pusa 10216 (GM chickpea) 108

qualified nondualism 160
quantitative trait loci (QTLs) in drought tolerant rice 108
quantum theory 103
see also interdependence
Quit India movement 43, 87

rajas 149, 151
Rajasthan 83, 124, 161-163
home to Bishnois and Bhils 161, 163
rakhees 39, 43
Ramayana 146, 149
and common *dharma* 152-153
Ramesh, Jairam 107
Rashtriya Swayamsevak Sangh (RSS) 30
see Hindu extremist groups
reforestation 161
regional economic development 50
reincarnation
Chipko's ecological 49
of the soul 169
relational worldview 103
religious environmentalism 143-144, 163, 167-168
Reni village 43, 48, 51
renunciation 145-146, 165
Resettlement and Rehabilitation (R&R) 72, 75, 122, 126
Review Committee on Genetic Manipulations (RCGM) 89
Ricetec control over rice 98-99
Right, universal principle of 148
right behaviour 27, 159
right to food 97-100
Rigveda 145, 157
Roy, Arundhati 66, 72
Roy, Raja Rammohan 147
Ruskin, John 27
Unto the Last 42, 168

Sahai, Dr. Suman 88
see also Gene Campaign (1993)
Samkhya 144, 146, 148-149
Sankat Mochan Foundation 164
santosha 153, 169
Sardar Sarovar Dam 65-66, 74, 128
Sardar Sarovar Narmada Nigam Limited 65
Sardar Sarovar Project (SSP)
as target of NBA protests 60, 66-67
cost-benefit issues 61, 63, 78
territorial conflicts 64-65
saucha 153, 169
Sarvodaya 29, 42, 45
and wellbeing of all creatures 96
three principles of 170
Sathya Sai Baba 161
Sathya Sai Central Trust 161

sattva 149-150
satyagraha
by NBA 66, 129
Gandhian concept 28
in Chipko movement 42, 55, 167
Navdanya's use of 96, 101
truth at all costs 119, 170
Saurashtra, Kutch, Rajasthan 124
scriptures, Hindu 166
influence on Gandhi 168-171
Seed Bill (2004) 85
seed freedom 95-96
seed *satyagraha* 96
seed sovereignty 92, 96, 105
Shankara 165
see also Advaita Vedanta
Shiva, Vandana 31
as founder of Navdanya 85
and Earth Democracy 103-105
Gandhian influence on 97
grassroots activism by 94
on Chipko movement 40, 50, 52
on climate changes and biodiversity 93-94
on 'food totalitarianism' 101
on violence arising from destruction of nature 103
opposition to GM crops 89, 98-101
Simon Company 42-43
slow food 101
Smritis 146
social inequalities 63, 125-126
social justice 32-33, 49, 56, 96-97, 109, 121, 125, 130, 136-137, 176
social transformation 50, 96, 105, 147, 165, 175
soil salinization 83
see also degradation
solidarity 44, 132
soteriology 149
Southern Materialist Ecofeminism 135-136
Sri Aurobindo 30, 147
Structural Adjustment Programme (SAP) 99
sugar cane crop 106
water usage of 107
GM alternative 106-107
Supreme Court of India 66-67, 78
sustainability 32, 49, 56, 78
and Hindu *dharma* 133-134, 153-154, 156-158
environmental 122, 129-130, 175-176
Gandhi's take on 118
of food supplies 98
of self to achieve *moksha* 156
see also Earth Democracy
sustainable agriculture 86, 88
as focus of NGOs and Gov't 92
as paradoxical in India 98
sustainable development
and gender equality 74
as a historical trend in Uttarakhand 40
definitions of 64
in coexistence with human rights 100
swadeshi 88, 96, 106
Jagran Manch (SJM) 106
swadharma 150
swadhyaya 153, 162
Swami Dayananda Saraswati 147
Swami Vivekananda 30, 147
swaraj 46, 96, 105, 116

tamas 149-151
tapasya 28
Tehri Garhwal 30, 37, 41, 43-44
temples 30, 129, 159, 161
Tennessee Valley Authority (TVA) 63
terminator technology 85
Tewari, K.M. 44
The Greater Common Good 72
Tenth Five Year Plan for 2002-2007 86
Three streams within Chipko movement 38
Tolstoy, Leo 170
Trade-related international property rights (TRIP) 86
transcendence 154-156, 160
transnational trade organizations 101
tribal populations 60, 128

'uncontrollable tenacious otherness' 132
UN Committee on Economic, Social and Cultural Rights (1999) 99
UN conference on the Human Environment (1972) 72
universal code 158
Universal Declaration of Human Rights 100
University of Delhi 107
universalizability principle 147-148, 153
Upanishads 144-145, 159
Uttar Pradesh Resin and Forest Produce Act of 1975 42
Uttarakhand 34
brief history of 41, 49-50
deforestation of 38
diet of the people 40
emigration from 41
local deities of 40
see also Chamoli Workers' Cooperative
Uttarakhand Sangharsh Vahini 54
Uttaranchal state 49-52

Vaisheshika 144
van panchayat 26, 42
rules revised (1976) 42
women taking control 47-48
Vandana Shiva
analysis of Chipko movement 150-152
see also Shiva, Vandana
varnadharma 150
Vasantdada Sugar Institute (VSI) 106
Vasudhaib Kutumbakam 31, 103

Vedanta 145-148, 164
Vedas 144-145
Vedic texts 145-157
'veteran intellectuals' in Navdanya's Garwhal study 94
Vinoba Bhave 42
virtue ethics 180
vishistadvaita 160, 165
Vishwa Hindu Parishad (VHP) 30
  also see Hindu extremist groups
Vivekananda, Swami 30, 44, 147
Vrindavan 165

water sovereignty 105
way of action (*karma yoga*) 150, 157, 162
way of devotion (*bhakti yoga*) 150, 169
way of knowledge (*jnana yoga*) 150, 157, 169
well-being 170-179
Western imperialism 23
Western moral philosophy 148
Who Really Feeds the World (2016) 103
  see also Shiva, Vandana
Women's impact on Chipko Movement 47-48
World Bank 61, 65, 67-68, 76-78, 99, 101, 121
  and profits from agribusiness 99
  disregarding cultural ties 129
  funding by 61, 65, 66-68, 130
  Inspection Panel 68
  lacking morality 155
  Morse Report 67
World Commission on Dams 62, 68
World Commission on Environment and Development 64
World Conservation Union 68
World Food Summit 90
World Trade Organization (WTO) 85, 89, 96

Yamuna River 161, 164-165
Yang, Tongjin 21
Yogasutra 149, 153, 168
Young India 28-29, 116

For Product Safety Concerns and Information please contact our EU representative GPSR@taylorandfrancis.com
Taylor & Francis Verlag GmbH, Kaufingerstraße 24, 80331 München, Germany

www.ingramcontent.com/pod-product-compliance
Ingram Content Group UK Ltd.
Pitfield, Milton Keynes, MK11 3LW, UK
UKHW011015180926
475701UK00004B/746
* 9 7 8 1 0 4 1 1 7 8 8 2 8 *